Water Resources Allocation

SHARING RISKS AND OPPORTUNITIES

This work is published under the responsibility of the Secretary-General of the OECD. The opinions expressed and arguments employed herein do not necessarily reflect the official views of OECD member countries.

This document and any map included herein are without prejudice to the status of or sovereignty over any territory, to the delimitation of international frontiers and boundaries and to the name of any territory, city or area.

Please cite this publication as:
OECD (2015), *Water Resources Allocation: Sharing Risks and Opportunities*, OECD Studies on Water, OECD Publishing, Paris.
http://dx.doi.org/10.1787/9789264229631-en

ISBN 978-92-64-22962-4 (print)
ISBN 978-92-64-22963-1 (PDF)

Series: OECD Studies on Water
ISSN 2224-5073 (print)
ISSN 2224-5081 (online)

Co-published by IWA Publishing
Alliance House, 12 Caxton Street, London SW1H 0QS, UK
Tel: +44 (0)20 7654 5500, Fax: +44 (0)20 7654 5555
publications@iwap.co.uk
www.iwapublishing.com

ISBN: 9781780407616 (Paperback)
ISBN: 9781780407623 (eBook)

The statistical data for Israel are supplied by and under the responsibility of the relevant Israeli authorities. The use of such data by the OECD is without prejudice to the status of the Golan Heights, East Jerusalem and Israeli settlements in the West Bank under the terms of international law.

Photo credits: Cover © Taro Yamada/Corbis; © iofoto/Shutterstock.com.

Foreword

The intensifying competition for water resources is widely documented. The OECD Environmental Outlook to 2050 highlights that water resources are already over-used or over-allocated in many places, with global demand expected to increase by 55% between 2000 and mid-century. The situation is compounded by climate change, with impacts on water expected to become more pronounced in the coming decades. These pressures have already made water allocation an urgent issue in a number of countries and one that is rising on the agenda in many others. Within this context, the OECD undertook work on water resources allocation to strengthen the evidence base and develop policy guidance to improve the design of allocation regimes and manage the challenges of reform.

This report is an output of the OECD Environment Directorate, under the leadership of Director Simon Upton. The project co-ordinator and lead author of the report is Kathleen Dominique. The Head of the Water Unit in the Environment Directorate, Xavier Leflaive, provided substantive inputs and advice. The OECD Secretariat project team, including Kun Wook Kim and Ignacio Deregibus, provided substantive inputs and analysis, in particular in the development of the Survey of Water Resources Allocation and the case studies of reform. The project was undertaken in collaboration with the Stockholm International Water Institute (SIWI) which provided in-kind support and expertise. The author is particularly grateful to John Joyce, Chief Economist at SIWI, for his sustained engagement and valuable insights.

Case studies on the reform of water allocation regimes were prepared by the OECD Secretariat project team, as well as Ian Barker of Water Policy International, Chibesa Pensulo and Johanna Sjödin of SIWI, Barbara Schreiner of Pegasys Strategy and Development, and Gavin Quibell, independent consultant. A background paper prepared Professor Mike Young, Chair in Water and Environmental Policy at the University of Adelaide provided a solid foundation for the work. The financial and in-kind contributions of the government of the Netherlands and the government of Korea helped to make this work possible and are greatly appreciated.

The project benefitted from the discussions at two international workshops supported by the Netherlands. The first, held in November 2012 in Wageningen, helped to frame the project. The second, held in May 2014 in The Hague, focused on the analytical framework and the policy guidance reflected in the "Health Check" for Water Resources Allocation.

The author is also grateful to colleagues and experts who provided valuable comments, including Simon Upton, Anthony Cox, Jane Ellis, Guillaume Gruère, Jesus Anton, Julien Hardelin and Delphine Clavreul (OECD Secretariat) and prominent water experts Ian Barker, of Water Policy International, John Matthews, of the Alliance for Global Water Adaptation, and Professor Mike Young. The contributions of the delegates of the OECD Working Party on Biodiversity, Water, and Ecosystems were instrumental in building a solid evidence basis for the analysis and in shaping the "Health Check". Keen editorial guidance from Janine Treves and impeccable administrative support from Sama Al Taher Cucci are also gratefully acknowledged.

Table of contents

Preface

Competition to access water resources is increasing as a result of population growth, economic development and climate change. As such competition intensifies, the issue of how governments allocate water between uses and users is rising on the policy agenda. International best practice shows that well-designed water allocation regimes help allocate water to where it creates the most value (economic, ecological, or socio-cultural) for society. They can adjust to changing conditions and preferences at least cost for society and can provide incentives for investment in water use efficiency and innovation.

Yet, in most countries today, rules and priorities for water allocation often do not reflect best practice. In many cases, these rules have developed over decades, or even centuries, and tend to be outdated or not adjusted to take into account new needs and realities. Growing pressures are making existing inefficiencies in water allocation regimes increasingly costly: 19th century allocation arrangements are poorly equipped to serve a 21st century society and economy. Conflicts over water use have an impact on both economic growth and environmental sustainability. However, allocation regimes have proven hard to adjust, even as the economic and social values related to water use have shifted over time. This means that water is often locked-in to uses that are no longer as valuable today as they were decades ago and that the risk of shortage falls disproportionately on certain groups of users. Allocation regimes are often ill-prepared to face more rapid and pronounced change in the future, especially in relation to climate change.

This report, *Water Resources Allocation: Sharing Risks and Opportunities*, takes a major step forward in providing policy guidance for countries seeking opportunities to unlock the value of water resources and to navigate the challenges of water allocation reform. A survey across 27 OECD and key partner countries for the first time establishes a solid evidence base of the current water allocation landscape. It reveals that most allocation regimes have elements that can encourage a robust system, but operate with significant limitations. For example, many allocation regimes suffer from legal ambiguities and unsustainable abstraction levels. Moreover, many countries still apply very low or no charges at all for water abstraction, even though the value of the resource has grown as competition for the resource has intensified.

While the case for reforming water allocation in many countries is clear, how to navigate the transition is particularly challenging. This report draws lessons from the reform experiences of 10 countries to address questions such as: How can intelligent sequencing help the reform process? How to compensate losers? And how to balance competing interests and objectives? The report also provides a practical tool that can be used to undertake a periodic "health check" of current allocation arrangements and identify opportunities for improving performance.

Although reforming water allocation may appear daunting, an improved regime can greatly increase the value that individuals and society obtain from water resources today and in the future. I am confident that policy makers can find both inspiration and pragmatic support in this report.

Angel Gurría,
OECD Secretary-General

Acronyms

AWRM	Active Water Resources Management
DEFRA	Department for Environment, Food and Rural Affairs, the United Kingdom
DGA	Directorate General of Water, Chile
DWA	Department of Water Affairs, South Africa
EEA	European Environment Agency
EIA	Environmental Impact Assessment
ELOHA	Ecological limits of hydrologic alteration
EU	European Union
FAO	Food and Agriculture Organization of the United Nations
LEMA	Law on Water and the Aquatic Environment, France (*Loi sur l'eau et les milieux aquatiques*)
MCM/Y	Million cubic meters per year
OUCG	Single Collective Management Bodies (*Organismes uniques de gestion collective*)
RBMP	River Basin Management Plan
UKCP09	United Kingdom Climate Projections
UNECE	United Nations Economic Commission for Europe
WAAs	Water apportionment areas (*zones de répartition des eaux*)
WFD	EU Water Framework Directive

Executive summary

Water resources allocation determines who is able to use water resources, how, when and where. Findings from a recent OECD survey of the current allocation landscape across 27 OECD and key partner countries, a first of its kind, reveal that most allocation regimes have elements that can encourage a robust system, but operate with significant limitations. Most allocation regimes today are strongly conditioned by historical preferences and usage patterns, tracing their roots to previous decades or even centuries. They have often evolved in a piecemeal fashion over time and exhibit a high degree of path dependency, which manifests in laws and policies, and even in the design and operational rules of long-lived water infrastructures. This means that water use is often "locked-in" to uses that are no longer as valuable today as they were decades ago, curtailing the value (ecological, socio-cultural, or economic) that individuals and society obtain from water.

In essence, allocation is a means to manage the risk of shortage and to adjudicate between competing uses. Allocation arrangements consist of a combination of policies, laws, and mechanisms. Water is a complex resource, with distinctive features as an economic good usually enjoying a distinctive legal status managed under the Public Trust doctrine. Access to the resource is usually subject to usage rights (or "water entitlements"), rather than outright ownership of the resource (with a few exceptions, in particular in the case of groundwater).

Strongly shaped by historical preferences that have proved difficult to change, existing allocation regimes are usually not well-equipped to deal with mounting pressure on the resource from intensifying competition, climate change, or shifts in societal preferences, such as increasing value placed on water-related ecological services. In order to reap greater benefits from water resources, an allocation regime needs to have two key characteristics: it should be robust by performing well under both average and extreme conditions and demonstrate adaptive efficiency in order to adjust to changing conditions at least cost over time. The challenges for allocation regimes are aggravated by the entrenchment of weak water policies (under-pricing water or an absence of regulation on use), which contributes to structural water scarcity, increasing the risk of shortage for users and for the environment.

The results of the OECD survey of the current allocation landscape indicate that the building blocks of a robust regime already exist in many cases, but their design has significant limitations. For example, there can be ambiguity between various legal regimes governing access to water resources (e.g. customary rights versus rights designated in formal laws). This legal "pluralism" is a source of conflict among water users and increases the likelihood of "allocation by litigation" or "allocation by adjudication", a costly and time-consuming process.

Adequate environmental flows are not secured in at least one-quarter of allocation regimes surveyed. Only 57% of allocation regimes surveyed report accounting for the potential impacts of climate change in their allocation arrangements, even though doing so is essential to ensure that allocation regimes can cope with changing conditions. Even less common are efforts to review shifting eco-hydrological baselines as climate conditions continue to alter the water cycle. A sequence of priority uses is clearly established in nearly all allocation regimes surveyed. This can be a practical approach to adjudicate among different types of users in times of shortage, but can also be the source of "lock-in", which can make it difficult to manage tensions among various users and place the risk of shortage disproportionately on "low priority" users. Water for the environment is rarely among the highest priorities in times of shortage and often figures among the lowest.

While nearly all of the examples surveyed (92%) reported having a clear definition of the limit (or "cap") on consumptive use, the cap may or may not be respected in practice. Only a few allocation regimes rely on flexible limits (as opposed to a fixed volume) defined in terms of the proportion of the resource that can be abstracted, thus reflecting variations in resource availability. There is clearly scope to broaden the application of abstraction charges. Given that abstraction charges tend to be low in most cases, increases in charges would improve cost recovery and provide a price signal to make low value and inefficient water uses less attractive. Most of the allocation regimes surveyed allow some form of trading, leasing or transferring water entitlements among users. However, a wide variety of conditions are placed on such transactions. These conditions can provide safeguards to minimise potential negative impacts, but also increase transaction costs and tend to limit the extent of trade that occurs in practice.

Recognising the potential for improving current allocation arrangements, 75% of countries covered in the OECD survey have recently reformed their allocation regimes and 62% have reforms ongoing. However, managing the transition from existing arrangements to an improved regime is often very contentious and can be costly. Evidence from case studies of allocation reform in 10 countries provides insight into the reform process and lessons on how some of the obstacles of reform can be overcome.

Concerns about water scarcity and insufficient water for ecosystems are often cited drivers of allocation reform. Broader political or structural reforms have provided imperatives to improve the efficiency of resource use and equity in allocation of water resources. Droughts can provide a salient, visible event to trigger action. The case studies on reform highlight the importance of determining a sustainable baseline (how much water is available for allocation) before making significant changes, like introducing trading. Failure to do so can result in costly efforts to claw back entitlements already granted. Willingness to engage stakeholders in the reform process and appropriately compensate potential "losers" (with financial transfers, permits to build storage structures) facilitates the process.

A periodic "health check" of current allocation arrangements can help to assess the achievement of reforms and areas for further improvement. The OECD *"Health Check" for Water Resources Allocation* can provide useful guidance for such a review. It is a tool designed to review current allocation arrangements to check whether the elements of a well-designed allocation regime are in place and to identify areas for potential improvement. In general, as the risk of shortage increases, the benefits of a more elaborate allocation regime increases. In the early stages of developing a water resource, or when the risk of shortage is low, a relatively simple allocation regime can be used with decisions made conservatively to avoid over-allocation and over-use. The basic building blocks of a robust regime should still be put into place, which can allow for adjustment at least cost over time as needed.

A well-designed allocation regime has multiple elements (discussed in detail in Chapter 5). A clear legal status should be in place for all types of water resources (surface and ground water, as well as alternative sources of supply) with competing claims clarified. A clear and enforceable abstraction limit ("cap") should be in place that accounts for *in situ* requirements and sustainable use, including environmental needs. Clearly defined, legal, volumetric water entitlements are needed. Water pricing, typically in the form of abstraction charges, is a key element of a well-designed regime. Pricing can contribute to cost recovery, internalise negative externalities associated with water abstractions, and send a price signal to users to discourage inefficient and low-value water uses. Scarcity pricing could help to signal the scarcity value of the resource, but has proven difficult to implement to date.

Although reforming an allocation regime can be a challenging process, an improved allocation regime can greatly increase the value (ecological, socio-cultural, or economic) that individuals and society obtain from water resources today and in the future. A periodic "health check" can provide a pragmatic approach to help realise these benefits.

"Health Check" for Water Resources Allocation

- **Check 1.** Are there accountability mechanisms in place for the management of water allocation that are effective at a catchment or basin scale?
- **Check 2.** Is there a clear legal status for all water resources (surface and ground water and alternative sources of supply)?
- **Check 3.** Is the availability of water resources (surface water, groundwater and alternative sources of supply) and possible scarcity well-understood?
- **Check 4.** Is there an abstraction limit ("cap") that reflects *in situ* requirements and sustainable use?
- **Check 5.** Is there an effective approach to enable efficient and fair management of the risk of shortage that ensures water for essential uses?
- **Check 6.** Are adequate arrangements in place for dealing with exceptional circumstances (such as drought or severe pollution events)?
- **Check 7.** Is there a process for dealing with new entrants and for increasing or varying existing entitlements?
- **Check 8.** Are there effective mechanisms for monitoring and enforcement, with clear and legally robust sanctions?
- **Check 9.** Are water infrastructures in place to store, treat and deliver water in order for the allocation regime to function effectively?
- **Check 10.** Is there policy coherence across sectors that affect water resources allocation?
- **Check 11.** Is there a clear legal definition of water entitlements?

- **Check 12.** Are appropriate abstraction charges in place for all users that reflect the impact of the abstraction on resource availability for other users and the environment?
- **Check 13.** Are obligations related to return flows and discharges properly specified and enforced?
- **Check 14.** Does the system allow water users to reallocate water among themselves to improve the allocative efficiency of the regime?

Chapter 1

Re-allocating water in a water scarce world

This chapter provides an overview of current and future challenges facing water allocation regimes, which are compromising their performance. It examines how competition for water resources is growing due to shifting demand, climate change, and changing societal preferences. The chapter also discusses how these pressures increase the value of well-designed allocation regimes that perform well across a range of conditions (averages as well as extremes) and can adapt to changing conditions at least cost.

Key messages

- Allocation regimes determine who is able to use water resources, how, when and where and directly affect the **value** (ecological, socio-cultural, or economic) that individuals and society obtain (or forego) from **water resources**.
- Allocation regimes are strongly conditioned by **historical preferences and usage patterns**. They trace their roots to previous decades or even centuries and have usually evolved in a piecemeal fashion.
- **Current and growing pressures** on water resources increase the **value of well-designed allocation regimes** that perform well across a range of conditions (averages as well as extremes) and can adapt to changing conditions at least cost.

Water resources serve multiple purposes and provide value to individuals, ecosystems, farms, firms, and society in various ways. The value obtained from water resources encompasses many forms – from ecological value provided by supporting key species, to socio-cultural value, to economic value derived from productive uses of water, to the existence value of iconic lakes or rivers.[1] How much water is left in water bodies (rivers, streams, aquifers) and how much is diverted for various uses; who is able to use these resources, how, when and where are questions that directly affect the value that individuals and society obtain from water resources. These questions are determined by allocation regimes, whether formal or informal. In this report, the term "allocation regime" is used to describe the combination of policies, mechanisms, and governance arrangements (entitlements, licenses, permits, etc.) used to determine who is allowed to abstract water from a resource pool, how much may be taken and when, as well as how much must be returned (of what quality), and the conditions associated with the use of this water (see glossary for key terms).[2]

The growing pressures on water resources and intensifying competition to access and use water is widely documented (OECD, 2012; WRI, 2015; UNESCO, 2012; Vörösmarty et al., 2010; Vörösmarty et al., 2002; Alcamo et al., 2000). Both demand and supply side pressures are on the rise, driven by economic development, population growth, deteriorating water quality and climate change. The *OECD Environmental Outlook to 2050* highlights that water resources are already over-used or over-allocated in many places. This is the case where current levels of abstraction exceed the sustainable level ("over-use") or where existing water entitlements (e.g. licenses or permits) to abstract water exceed the sustainable level ("over-allocation"). For example, groundwater[3] in many parts of the world is being exploited faster than it can be replenished and is also becoming increasingly degraded due to the impact of pollutants. Between 1960 and 2000, the rate of groundwater depletion more than doubled (OECD, 2012). Often, adequate environmental flows have not been secured, threatening the health of freshwater ecosystems.

The situation is compounded by climate change. Climate change increases water risks – both quantity and quality, along with disrupting freshwater ecosystems. It also generates increased uncertainty about future water availability and makes historical climate conditions a less reliable guide to current and future planning. Climate change can provoke significant shifts in the timing, location, amount and form of precipitation (for instance from snowfall to rain). More frequent and intense droughts can also be expected in many regions.[4] Beyond changes in both averages and extremes, climate change can also result in "state level" shifts. Increasing variability and less predictable supply pose new challenges for allocation (OECD, 2013a; OECD, 2014a).

These pressures have already made water allocation a pressing issue in a number of countries and an issue that is rising on the agenda in many others, even some traditionally water abundant countries, like Brazil, the Netherlands, France and the United Kingdom. Water allocation regimes in most countries have evolved gradually over time, often in a piecemeal fashion. Their design was driven by past development policies, based on historical water availability, and influenced by social preferences, general economic context and available techniques of previous generations. Some allocation regimes trace their roots to previous centuries and many were not designed to adjust to changing conditions and societal preferences. Growing pressures are making existing inefficiencies in water allocation regimes increasingly costly; 19th century allocation arrangements are poorly equipped to serve a 21st century society and economy.

However, once established, allocation arrangements have proven difficult, and often costly, to adjust. These allocation arrangements exhibit a high degree of path dependency, which manifests itself in both institutional arrangements (law, property rights and policies) and long-lived water infrastructures, such as dams, canals and pipelines. As a result, allocation regimes are usually not well-equipped to deal with mounting pressure on the resource, the emergence of new scientific understanding (of the resource or of related ecological needs), or adapt to shifts in societal preferences, such as increasing value placed on water-related ecological services. The challenges for allocation are aggravated by the entrenchment of weak water policies (under-pricing water or lack of regulating use), which contributes to structural water scarcity, increasing the risk of shortage for users and for the environment.

Growing pressures on water allocation regimes

The lack of a sufficient quantity of water of adequate quality to satisfy demand creates a risk of shortage for certain users. There are a number of drivers (both demand and supply side) that affect the risk of water shortage. The risk of shortage is determined by: 1) the consequences (impacts) of a shortage of water for a given use; and 2) the likelihood of its occurrence. It is driven by the intersection of hazards, exposure and vulnerability. Water shortage arises from conditions of scarcity, which can be defined as *"an imbalance between the supply and demand of freshwater as a result of a high level of demand compared to available supply, under prevailing institutional arrangements (including price) and infrastructural conditions"* (Winpenny, 2011).

When considering scarcity, it is useful to make a distinction between economic and absolute scarcity. Economic scarcity exists when there has been underinvestment in water infrastructure to supply sufficient amounts of water. Absolute scarcity exists when there is no affordable source of additional water, or where the costs of additional water supplies exceed the benefits of their provision. In the case of absolute scarcity, it is necessary to keep use within the limits of sustainable use. Once absolute scarcity has been reached, the design of the allocation regime becomes crucial (Young, 2013).

The availability of water is not only a question of quantity. Deteriorating water quality as a result of both point source and diffuse discharges, also affects water availability. Degraded water quality changes the economics of resources use, as inadequate quality requires treatment before use and reduces the value that can be derived from certain in-stream uses (bathing, ecosystem functioning, etc.).

Current and growing pressures that contribute to the risk of shortage increase the importance of well-designed allocation regimes. The availability of the resource pool for allocation is determined by the physical characteristics of water resources, investments in water infrastructure, as well as shifting, and increasingly less predictable climatic conditions. In a changing climate, the frequency and intensity of extreme events is projected to increase in many areas. With the "exceptional" becoming more commonplace, these changes mean that the way in which "exceptional circumstances" are currently defined in allocation regimes merit re-evaluation. Finally, changes in aggregate water demand, as well as the composition of that demand, affect the allocation among various uses, reducing or intensifying competition among and within certain categories of uses.

Changing societal preferences are an important factor in determining the repartition between *in situ* and diverted uses, in particular the determination of environmental flow requirements, and the sequence of priority uses (where they exist). Improvements in water use efficiency effectively change the rates of consumption, affecting return flows and the available resource pool. All of these trends create pressures on existing allocation regimes, increasing the value of flexible, clearly defined, effective and efficient allocation arrangements. These trends and their relevance for allocation are summarised in Table 1.1 and then discussed in the following sections.

Table 1.1. **Trends affecting water allocation regimes**

Trend	Implications for allocation regimes
Changing demand patterns	● Affects the available resource pool. ● Affects the competition among and within categories of water uses. ● Affects the type of water (piped, level of quality) demanded and the desired reliability of supply.
Climate change impacts on freshwater	● Shifts the timing, location, form and amount of precipitation, affecting the available resource pool. ● Increases uncertainty about the availability of freshwater. ● To the extent that extreme events will occur more frequently, makes "exceptional circumstances" more commonplace.
Deteriorating water quality	● Affects the available resource pool. ● Increases the cost of water resource use.
Improving water use efficiency	● Affects the available resource pool. ● Changes rates of consumption. ● May reduce return flows, reducing availability for subsequent uses. ● Increases the importance of specifying return flows in entitlements.
Shifting societal preferences	● Changes the value placed on *in situ* and diverted uses. ● Affects the value placed on water for environmental purposes, influencing the definition of environmental flows.
Improving scientific understanding of the resource or environmental flow requirements	● Affects the understanding of the resource and its interaction with other water bodies. ● Affects the definition of environmental flows or other *in situ* flow requirements. ● May also influence the understanding of the value of environmental flows (e.g. in terms of supporting key species or habitats).

Changing patterns of demand

A world economy four times larger in 2050, and with over 2 billion additional people, will need more water. Under the *OECD Environmental Outlook's* baseline scenario, global water demand is expected to increase by around 55% between 2000 and mid-century. This is primarily due to growing demand from manufacturing (+400%), thermal power plants (+140%) and domestic use (+130%), as depicted in Figure 1.1. As a result, there is little scope for increased use of irrigation water use in most regions. This "squeeze" on irrigation use comes about because domestic and industrial uses are usually prioritised over lower value or less efficient irrigation uses in water allocation regimes (at least in most OECD countries). Water for the environment will also be competing with these demands, adding to existing stressors on freshwater ecosystems (OECD, 2012).

However, trends in water demand diverge between OECD and non-OECD countries. In OECD countries, water demand is actually projected to decrease somewhat (from 1 000 km^3 in 2000 to 900 km^3 in 2050). This projected decrease in demand in OECD is expected to be driven by efficiency gains as well as a structural shift in the economy towards service sectors that are less water intensive. In contrast, water demand is projected to increase significantly in the BRIICS (from 1 900 km^3 in 2000 to 3 200 km^3 in 2050) and to a lesser extent in the rest of the world (from 700 km^3 in 2000 to 1 300 km^3 in 2050). Most of the population in river basins expected to be under severe water stress live in the BRIICS (OECD, 2012).

Figure 1.1. **Global water demand, baseline scenario from 2000-50**

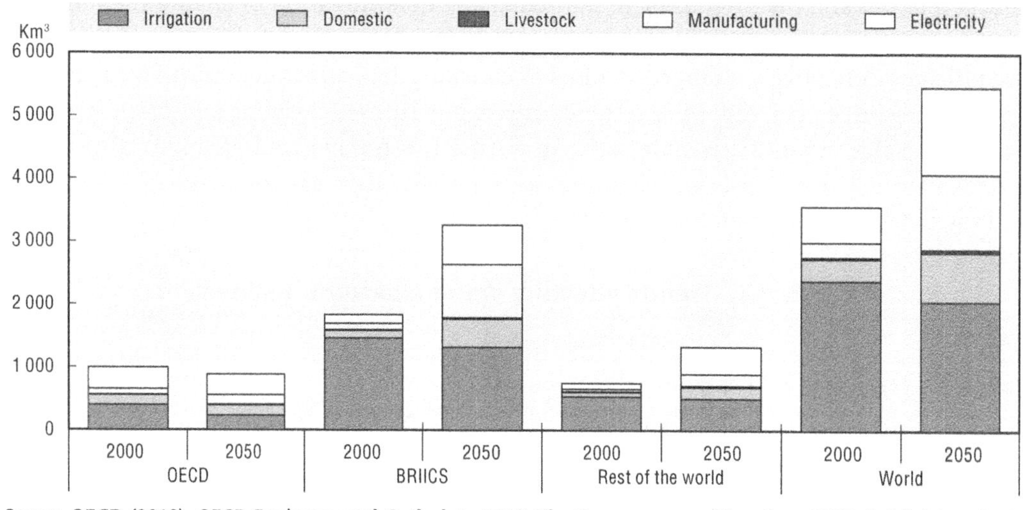

Source: OECD (2012), *OECD Environmental Outlook to 2050: The Consequences of Inaction*, OECD Publishing, Paris, *http://dx.doi.org/10.1787/9789264122246-en*; output from IMAGE.

Nevertheless, it is important to note that aggregate demand projections can be misleading, as figures at the national level can mask regional or local scarcity and temporal issues. The location and timing of demand relative to supply determines the scarcity conditions, and hence, the risk of shortage. Even water-abundant countries, like Brazil or the Netherlands, face localised and seasonal episodes of scarcity. As a result, it is doubtful, that this reduction in demand will be enough to address the serious regional stresses that already exist in parts of Australia, Israel, Mexico, Spain and the United States (OECD, 2013b) and emerging stresses elsewhere.

Changes in aggregate demand affect allocation, but how demand is repartitioned among various uses is also relevant. Recent trends in abstraction per use in OECD countries reveal a substantial re-allocation of water among various uses (Figure 1.2). Trends indicate a shift towards typically higher value and higher priority uses. For instance, the increase in

Figure 1.2. **Freshwater abstractions in OECD countries**

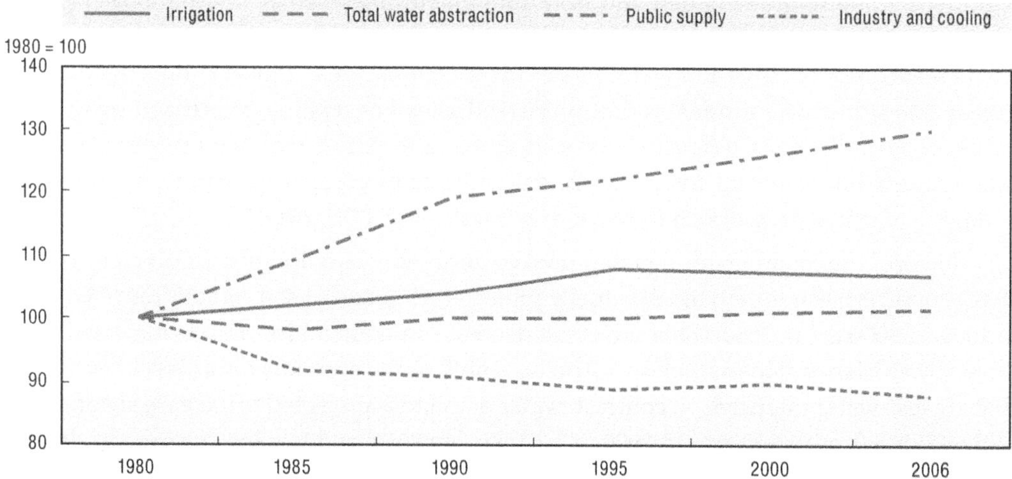

Source: OECD (2010), *OECD Factbook 2010: Economic, Environmental and Social Statistics*, OECD Publishing, Paris, *http://dx.doi.org/10.1787/factbook-2010-en*.

demand for public water supply (e.g. water supply for domestic consumption), shifting away from irrigation is significant since public supply uses tend to have a higher priority status in allocation regimes. There is also an implication for the type of water demanded (piped, level of quality) and the level of reliability required. While farmers growing annual crops can make adjustments in cropping decisions if rainfall is delayed, a fruit grower, a city or some industries cannot. If overall demand for water may be levelling off and even declining in OECD countries, competition among higher priority uses is in fact intensifying, narrowing the margin of manoeuvre for adjustment during episodes of shortage.

Climate change impacts on freshwater

In a changing climate, precipitation patterns are shifting rainy seasons and affecting the timing and quantity of melt water from snow pack and glaciers. More torrential rains, floods and droughts can be expected in many regions. Climate change impacts on freshwater resources are already evident and are projected to become more significant and to accelerate over time (Bates et al., 2008). Projected changes in the water cycle can have significant impacts on agricultural production in practically all regions of the world resulting in destabilising impacts for agricultural markets, food security and non-agricultural water uses (OECD, 2014a). These changes present a singular challenge for water systems by rendering the historical assumption of stationarity[5] increasingly unreliable as a basis for water management (Milly et al., 2008). This means that a fundamental assumption upon which many water allocation regimes are based will no longer be a sufficiently reliable basis for future planning and allocation (Brown, 2010).

Climate change is bringing about not just shifts in mean precipitation, but also shifts between seasons and between years, as well as extremes, with more frequent and severe floods and droughts expected in some regions. Changes may be gradual or sudden, resulting in "state level" shifts.[6]

Despite abundant evidence of climate change impacts on freshwater, there is significant uncertainty about the precise nature, timing and magnitude of expected shifts at the relevant scale for allocation decisions. The level of confidence in climate change projections for key water parameters decreases as their potential utility for water management decision-making increases (OECD, 2013a). In the future, allocation regimes will need to accommodate considerable uncertainty about water availability and the needs of ecosystems. Increasing variability and frequency of extremes as well as greater uncertainty about future conditions increases the value of well-designed and flexible allocation regimes.

One of the most common mistakes made when considering how best to manage water allocation is to assume that the impact of climate change on water supply will be gradual. Experience has shown that sudden climatic shifts can occur. In the case of Perth, a sudden shift appears to have occurred in 1974, as illustrated in Box 1.1 Since then, the amount of water available for consumptive use in this region has more than halved.

Reductions in rainfall can produce an even more drastic reduction in streamflow. In the case of Jarrahdale, a 14% reduction in rainfall resulted in 48% less stream inflow; 20% reduction in rainfall resulted in 66% less stream inflow. The impact of the reduction in stream inflow has an even greater impact on consumptive use. This is because sufficient base flows are still required before water can be extracted. This means that a relatively small reduction in mean rainfall can ultimately have a massive and disproportionate impact on the volume of water available for use (Figure 1.5).

Box 1.1. **Abrupt shifts to a drier climate, an example from Perth**

In 1974, the city of Perth in Western Australia appears to have experienced a sudden unexpected shift to a drier climatic pattern. Since that year, mean rainfall has been between 14% and 20% less than it was for the first two thirds of the century (Figure 1.3). As a result, inflows into Stirling Dam have more than halved (Figure 1.4).

This experience demonstrates how state level shifts in climatic patterns are possible and how relatively small reductions in mean rainfall can result in a dramatic reduction in the quantity of water available for use.

Figure 1.3. **Historical trends in rainfall for Jarrahdale**

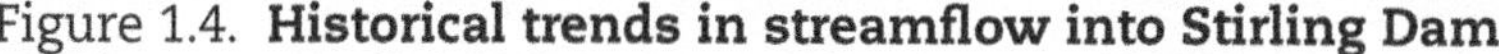

Figure 1.4. **Historical trends in streamflow into Stirling Dam**

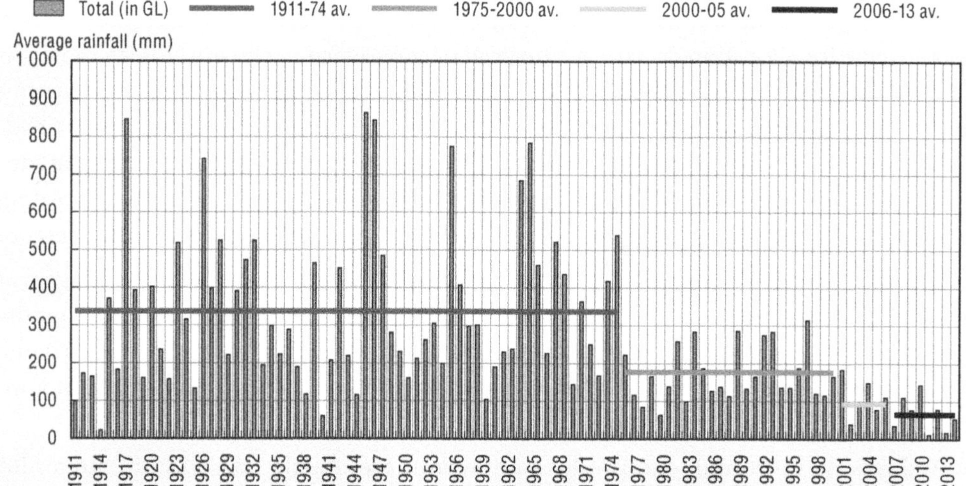

Notes: Streamflow is from May of labelled year to the following April. In order to provide an accurate historical comparison streamflow from Stirling, Wokalup and Samson Brook Dams are not included in this data as these dams only came online in 2001. Inflow is therefore modelled on Perth dams pre-2001.

Source: Water Corporation of Western Australia (2014), *www.watercorporation.com.au/water-supply-and-services/rainfall-and-dams/streamflow/streamflowhistorical.*

Figure 1.5. **Effect of reduction of stream inflow on the amount of water available for consumptive use**

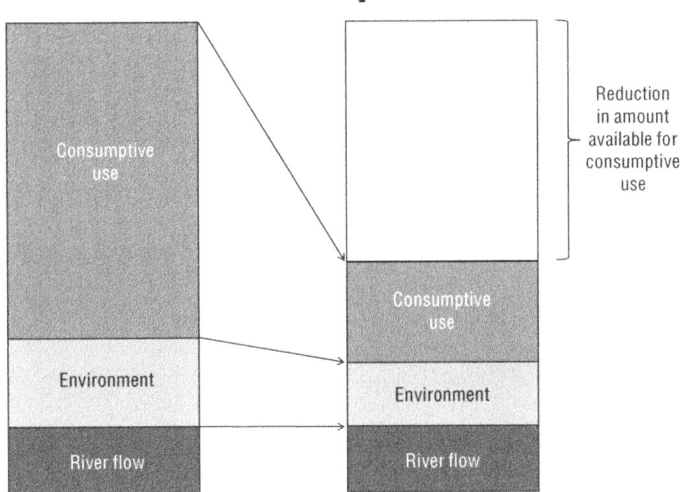

Source: Young, M. (2013), *Improving Water Entitlement and Allocation*, background paper for the OECD project on water resources allocation (unpublished).

Deteriorating water quality

Water availability is affected not only by the available quantity of water, but also its quality. Deteriorating water quality constrains water availability (varying by type of use and the degree of quality required). It also affects the economics of water resource use, as poor water quality requires costly treatment before use and also reduces the value that can be derived from in-stream uses (impeding ecosystem functioning, fisheries, bathing, etc.).

Globally, trends indicate that water quality is expected to stabilise or improve in most OECD countries by 2050, while outside the OECD, water quality is expected to deteriorate (OECD, 2012). Declining water quality is mainly due to nutrient flows from agriculture and from absent or poor wastewater treatment. This results in increased eutrophication, biodiversity loss, water-related disease and an increase in costs for treatment prior to use. Worldwide, a significant portion of wastewater remains untreated, especially in developing countries. Water pollution from urban sewage is expected to increase 3-fold by 2050, as compared to 2000 (Figure 1.6), as progress in urbanisation and wastewater collection outpace investment in wastewater treatment.

Another example of deteriorating water quality is in the Netherlands, where increasing salinity in some regions has contributed to the increasing risk of shortage. This is spurring a review of policy options for freshwater supply, including allocation arrangements (Box 1.2).

Water use efficiency gains and changes in rates of water consumption

Improvement in water use efficiency is among the factors behind declining demand in OECD countries. Generally, this is a positive trend, as improved efficiency can relieve stress on water resources and free up water for other uses (*in situ* or diverted). However, even radical gains in efficiency of current uses may not be enough to avoid a more fundamental appraisal of the allocation of water (OECD, 2012). Furthermore, efficiency improvements can result in unintended consequences for water allocation, when consumption rates change and return flows are not properly accounted for.

Figure 1.6. **Nitrogen effluents from wastewater: 2000 to 2050**

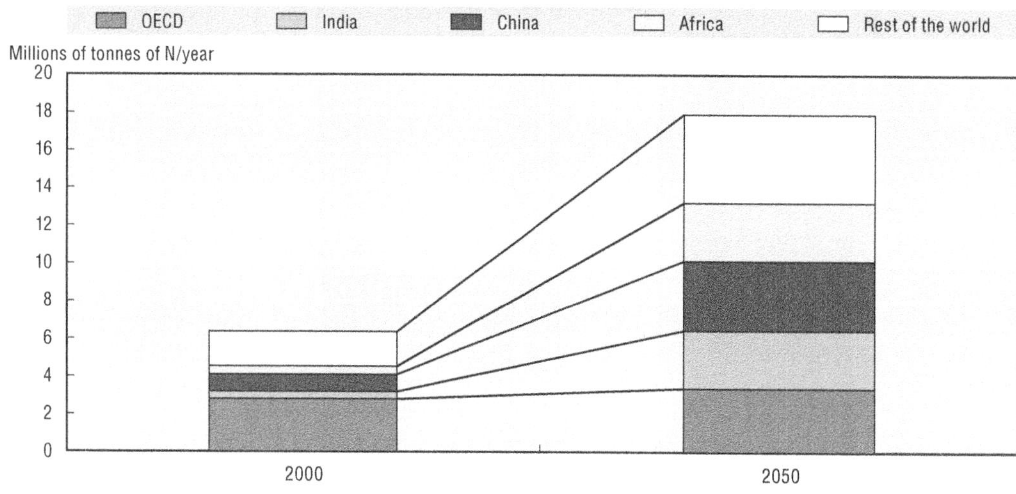

Source: OECD (2012), *OECD Environmental Outlook to 2050: The Consequences of Inaction*, OECD Publishing, Paris, http://dx.doi.org/10.1787/9789264122246-en.

While some arid countries have been managing water scarcity for centuries, in many other countries water allocation regimes were established during times of abundance (perceived or actual). As a result, it is common for many regimes to give little attention to the effects of use by one user on the use of another user. One of the most common oversights is a failure to account for return flows. Thus, in many allocation regimes, return flows of water entitlements are not specified, and consumption rates of various uses are estimated (if at all) using generic co-efficients. Increasing efficiency of water use increases the consumption-intensity, meaning that return flows (a positive externality of water use) are reduced. This leaves less water available to seep into groundwater or available for downstream users. This situation can also arise with significant interceptions of run-off, such as afforestation, which are not usually considered "water users" in formal arrangements. This can result in an over-estimation of available resources and as a result, over-allocated or over-used resources.

A striking example is the case of water use for energy production. According to the IEA's *World Energy Outlook* (2012), the water-intensity of global withdrawals and consumption for energy production (water withdrawals and consumption per unit of energy produced) head in opposite directions in the period assessed in the study. Figure 1.7 depicts projected shifts in water-intensity of energy production. Withdrawal-intensity of global energy production falls by 23%, whereas consumption-intensity increases by almost 18%. This is mainly the result of an expected shift in the power sector away from traditional once-through cooling systems towards wet towers (that reduce withdrawals but raise consumption).

A reduction in withdrawal-intensity implies that less water will be abstracted per unit of energy produced, which can have a positive impact on water resources allocation (depending on the overall trend for energy production) by freeing up water for other uses. However, as consumption-intensity increases and if return flows are not properly accounted for, the integrity of the allocation system may be undermined.

Box 1.2. **Increasing risk of shortage in the Netherlands strains the current approach to water allocation**

Even in a water abundant country like the Netherlands, periodic and localised scarcity can arise, putting pressure on the existing allocation regime and resulting in costly impacts. The Netherlands is experiencing a growing risk of shortage due to a lack of water in some regions and increasing salinity in others as sea water intrudes into the delta and saline groundwater rises. Climate change is expected to increase the variability of water supply and slightly reduce water availability. Periods of drought and low river discharge occurred in 1976, during the very dry summer of 2003, the dry spring of 2005 and in 2011. In some cases, water had to be trucked in from other regions to avoid significant losses to high value agriculture. Competition among users is also intensifying, such as increasing demand for electricity, increasing power stations, which sometimes conflicts with other water interests (OECD, 2014b).

Water shortages impose significant costs or result in reduced revenues or benefits for the agriculture, shipping and energy sectors, and for nature conservation and recreational uses. Remarkably, with a total surface area of 41 500 km^2 (of which 7 500 km^2 is water and 19 100 km^2 is agricultural area), the Netherlands is one of the largest net exporters of agricultural products and foods in the world (along with France and the United States), exporting EUR 65 billion worth of vegetables, fruit, flowers, meat and dairy products each year (OECD, 2014b). According to Jeuken et al. (2012), estimates of economic loss to the Dutch agricultural sector may reach EUR 700 million in a "dry year" (frequency of 1 out of 10 years) and EUR 1 800 million in an "extreme dry year" (frequency of 1 out of 100 years).[*] These figures are equal to 0.1% and 0.3% of GDP respectively. These damages could increase significantly due to climate change and socio-economic developments. The Ministry of Economy, of Economic Affairs (2011) estimated that damages could increase fivefold in 2050, translating into a loss for the agricultural sector of EUR 700 million once every two years (Jeuken et al., 2012).

Recognition of the increasing risk of shortage and the challenges they present, the Delta Programme has included freshwater supply as one of the main programme areas. It is spurring reconsideration of the prevailing approach to allocation. Currently, a priority regime is used to limit abstractions during periods of water shortage. Flood safety (ensuring that dykes do not dry out and collapse) and the prevention of irreversible damage to the environment are the top priority. Second priority is given to drinking water and power supply needs. Capital-intensive agriculture and industrial uses are the third priority, while the fourth priority is given to other types of agriculture, the environment (aside from cases where irreversible damage can occur) and other uses.

Managing shortage incidents takes the form of priority regime banning. This is a pragmatic approach and works as a short-term strategy in cases where shortage incidents occur infrequently. It also has a relatively low cost of implementation as elaborate administrative arrangements are not necessary. However, the blunt nature of priority ranking regimes means that there are few incentives for water users to proactively manage the risk of shortage. When shortage occurs in such a priority regime, the introduction of a ban is often sudden and final. Users within a given priority category are treated uniformly, even if there are significant differences in their water needs, the value they assign to water, or their risk preferences (e.g. willingness to pay to avoid the risk of shortage). They have few or no options to respond when shortage occurs. Since the expectation is that water shortages will become more frequent in the future, the limitations of the current approach are likely to become more evident.

[*] A "dry year" has a precipitation deficiency of more than 220 mm in the summer. An "extreme dry year" has a precipitation deficiency of over 360 mm in the summer.
Sources: OECD (2014b); Jeuken et al. (2012).

Figure 1.7. **Projected shifts in water-intensity of energy production**

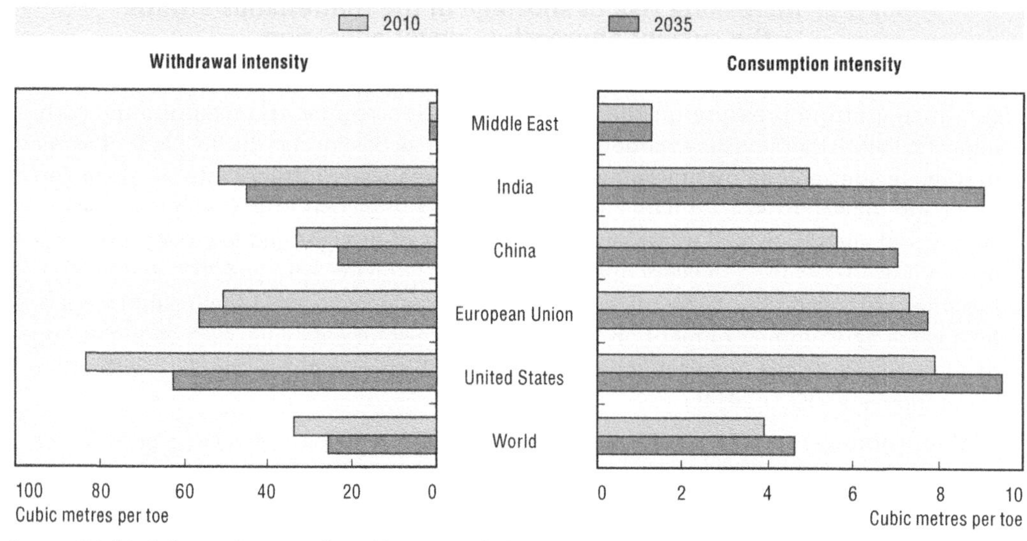

Source: IEA (2012), "Water for Energy", World Energy Outlook 2012, OECD Publishing, Paris, http://dx.doi.org/10.1787/weo-2012-en.

Shifting social preferences

The value derived from water uses and the objectives of water allocation have changed over time, reflecting shifting social preferences. Water resources have traditionally provided a range of socio-cultural values. These include the existence value of an iconic lake or river; aesthetic value, which may be reflected in property values located near attractive water bodies; or values associated with Indigenous heritage or other community uses of water resources. Historically, the terms of access to water have often been used as an incentive to drive certain development priorities, including supporting irrigated agriculture, energy production and water-intensive industrial sectors. In the western United States, water use entitlements were used as an incentive to settle and develop land, privileging early users over those who made claims later. This arrangement was institutionalised in the system of *prior appropriation*, whereby the "first in time" is the "first in line" to access water, which has proven difficult to change.

Water requirements to ensure adequate environmental flows have begun to be seen as a legitimate and valuable use of water only relatively recently in many countries. The need to restore adequate environmental flows to support vital ecosystem functions and freshwater biodiversity is generating increasing attention in many countries. This is reflected in efforts to allocate more water to support ecosystem services, often requiring the curtailment of diverted uses in situations of scarcity. The EU Water Framework Directive's requirement for member states to achieve good ecological status, for which flow is a supporting factor, for natural water bodies, (and good ecological *potential* for heavily modified and man-made water bodies) is indicative of this trend. The need to secure sufficient water for the environment is among the factors driving the reform of the abstraction licensing regime in England and Wales (Box 1.3).

Another example of this trend is seen in Australia. In the Murray-Darling Basin, an environmental watering plan has been developed to co-ordinate environment water use. One of the three primary objectives of the draft plan is ensure the resilience of water-related ecosystems to climate change and other risks. To help achieve these aims, the

---

Box 1.3. Inflexibility of current regime and growing pressures help build the case for abstraction reform in England and Wales

Despite its reputation as a wet country, the east and southeast of England have very low water resource availability, with as little as 300 mm per annum of effective rainfall. Many rivers are of high ecological sensitivity and of international importance for their conservation value. Population density is highest in these areas of lowest rainfall so demand is greatest where resources are the scarcest. This has led to many rivers are being damaged or threatened by unsustainable abstraction.

The current system for managing abstraction of water from rivers and aquifers in England was set up in the 1960s, when water was perceived to be abundant. It was not originally intended to manage competing demands for water. Passed in 1963, the Water Resources Act was, for its time, innovative and far-reaching. Although the act required every licence to be assessed according to its reasonable need and its impact on the aquatic environment, over time the former has changed and the latter is now much better understood. In addition, changing patterns of demand have left many licences under-utilised, with no straight-forward mechanism for trading resources.

The United Kingdom's Environment Agency manages and regulates the allocation of water resources in all catchments in England. It currently has a major programme to address these damaging abstractions. However, because abstraction licences are deemed to be a property right, compensation may be payable when licences are forcibly changed or revoked. This means that the process of achieving a sustainable flow regime, or reacting to changing patterns of resource availability or demand, can be slow and expensive.

The Environment Agency has modelled a range of climate change scenarios to understand their potential impact on river flows and water availability out to the 2050s. It has also matched these with scenario planning using different socio-economic models in order to understand how demand for different purposes (public water supply, energy and agriculture) might change over a similar time horizon. This work has demonstrated that under some scenarios, water availability could decrease significantly and that demand – driven in particular by projected population growth – might also increase in a way which would mean widespread impacts on rivers and ecosystems together with risks to security of supplies.

Although there is a structured approach to public water supply planning which takes account of changes in demand and the impacts of climate change over a 25-year horizon, the Environment Agency and the United Kingdom government were concerned about the long-term risks to the environment and water supplies.* It was recognised that the current system is too inflexible to be able to cope adequately with changes in demand and resource availability, and potentially could act as a drag on economic growth. The government launched a consultation in December 2013 on proposals for a more flexible and dynamic system which would be able to react to future uncertainties and allow access to resources in a reformed regulatory system. At its heart, any new system would ensure that there was sufficient water for the environment, adequately protected at all flow states, and that above this threshold, water would be available for allocation. As long as all abstraction licences have a sustainable basis, there is then the potential for greater trading and economic benefit from more efficient use.

In parallel, water companies are responding to the need to improve the connectivity of their supply systems in order to increase their resilience, and also to seek opportunities for sharing water across company boundaries. The Environment Agency is also working to take a more strategic approach to the long-term water demands of the agriculture sector and energy generation so as to drive a more integrated approach to resource management across the water-food-energy nexus.

* The Environment Agency developed a "Case for Change" which supported a government *White Paper* ("Water for Life") in 2011. This set out a range of proposals for the reform of the public water supply industry and the abstraction licensing system.

Source: OECD (2014b) based on input from Ian Barker, former Head of Water, Land and Biodiversity at the UK Environment Agency.

Commonwealth Environmental Water Holder was established as a statutory position under the Water Act to manage water recovered by the Australian Government to protect and restore environmental assets. A water entitlement buy-back programme is underway, with AUD 3.1 billion set aside for purchasing water entitlements to help restore the health of vitally important rivers, wetlands and floodplains (OECD, 2013a). These policy orientations reflect the greater value placed by society on ensuring adequate water to maintain ecosystem services. They also require a re-evaluation of allocation arrangements.

Conclusion

Growing pressures are making existing inefficiencies in water allocation regimes increasingly costly. Current water allocation regimes are strongly shaped by historical preferences and usage patterns. Most allocation regimes have been established in piecemeal fashion over the course of previous decades (and sometimes centuries). As such, they are usually poorly equipped to adjust to changing conditions and to deal efficiently with current and future pressures. This can result in various costs, or lost opportunities to capture greater value from water resources. Impacts include degraded environmental performance (where minimal flows required to support ecosystem services are not secured), lost opportunities for economic development (when water insecurity holds back investment), and inequitable management of water risks (where responses to water scarcity, such as banning low priority uses, place the risk of shortage disproportionally on certain groups of users).

To ensure that the use of water resources can better meet the needs of society today and in the future, allocation regimes deserve review and where necessary, reform. The remainder of the report elaborates a framework that public authorities and water users interested in identifying and improving allocation arrangements can use to enhance their performance and reap greater benefits from water resources.

Notes

1. The TEEB report, *The Economics of Ecosystems and Biodiversity Ecological and Economic Foundations* (2010), provides a detailed discussion on the various types of values provided by natural capital.

2. This report discusses issues related to both surface and groundwater allocation. The specificities of groundwater allocation will examined in greater depth in a forthcoming OECD project. See: OECD (2015a) for a report on groundwater use in agriculture in OECD countries.

3. See: OECD (2015a) for a comprehensive overview of the status and characterisation of groundwater resources in agriculture in OECD countries.

4. See: OECD (2015b) for a comprehensive analysis of policy approaches to droughts and floods in agriculture.

5. As expressed in Milly et al. (2008), stationarity is the idea that natural systems fluctuate within an unchanging envelope of variability. It is a foundational concept that permeates training and practice in water-resource engineering.

6. See: Wickel and Matthews (2012), "Modes of Climate Change", in OECD (2013), *Water and Climate Change Adaptation: Policies to Navigate Uncharted Waters*, OECD Publishing, Paris.

References

Alcamo, J. et al. (2000), "World Water in 2025 – Global Modelling and Scenario Analysis for the World Commission on Water for the 21st Century", *Report A0002*, Centre for Environmental Systems Research, University of Kassel.

Bates, B. et al. (eds.) (2008), "Climate Change and Water", *Technical Paper VI*, Intergovernmental Panel on Climate Change (IPCC) Secretariat, Geneva.

Brown, C. (2010), "The End of Reliability", *Journal of Water Resources Planning and Management*, Vol. 136, American Society of Civil Engineers, pp. 143-145.

DEFRA (2011), *Water for Life*, Department for Environment Food and Rural Affairs, HM Government, *www.gov.uk/government/uploads/system/uploads/attachment_data/file/228861/8230.pdf* (accessed 10 January 2014).

Grafton, R.Q. et al. (2012), "A Comparative Assessment of Water Markets: Insights from the Murray-Darling Basin of Australia and the Western US", *Water Policy*, Vol. 14, No. 2.

IEA (2012), "Water for Energy", *World Energy Outlook 2012*, OECD Publishing, Paris, *http://dx.doi.org/10.1787/weo-2012-en*.

Jeuken, A. et al. (2012), "Balancing supply and demand of freshwater under increasing drought and salinisation in the Netherlands", *Midterm Report Knowledge for Climate Theme 2*, KfC Report, No. 58/2012.

Klijn, F., E. van Velzen and J. Hunink (2012), *Zoetwatervoorziening in Nederland aangescherpte landelijkeKnelpuntenanalyse*, Deltares.

Milly, P.C.D. et al. (2008), "Stationarity is Dead: Whither Water Management", *Science*, Vol. 319, American Association for the Advancement of Science, pp. 573-574, *http://dx.doi.org/10.1126/science.1151915*.

National Water Commission (2012), "The Impacts of Water Trading in the Southern Murray-Darling Basin between 2006-07 and 2010-11", National Water Commission, Canberra.

National Water Commission (2010), *The Impacts of Water Trading in the Southern Murray-Darling Basin: An Economic, Social and Environmental Assessment*, National Water Commission, Canberra.

OECD (2015a), *Groundwater Use in Agriculture in OECD Countries: From Economic Challenges to Policy Solutions*, OECD Publishing, Paris, forthcoming.

OECD (2015b), *Policy Approaches to Droughts and Floods in Agriculture*, OECD Publishing, Paris, forthcoming.

OECD (2014a), *Climate Change, Water and Agriculture: Towards Resilient Systems*, OECD Studies on Water, OECD Publishing, Paris, *http://dx.doi.org/10.1787/9789264209138-en*.

OECD (2014b), *Water Governance in the Netherlands: Fit for the Future?*, OECD Studies on Water, OECD Publishing, Paris, *http://dx.doi.org/10.1787/9789264102637-en*.

OECD (2013a), *Water and Climate Change Adaptation: Policies to Navigate Uncharted Waters*, OECD Studies on Water, OECD Publishing, Paris, *http://dx.doi.org/10.1787/9789264200449-en*.

OECD (2013b), *Water Security for Better Lives*, OECD Studies on Water, OECD Publishing, Paris, *http://dx.doi.org/10.1787/9789264202405-en*.

OECD (2012), *OECD Environmental Outlook to 2050: The Consequences of Inaction*, OECD Publishing, Paris, *http://dx.doi.org/10.1787/9789264122246-en*.

OECD (2010), *OECD Factbook 2010: Economic, Environmental and Social Statistics*, OECD Publishing, Paris, *http://dx.doi.org/10.1787/factbook-2010-en*.

TEEB (2010), *The Economics of Ecosystems and Biodiversity (TEEB) Ecological and Economic Foundations*, in Kumar, P. (ed.), Earthscan, London and Washington, *www.teebweb.org/publication/the-economics-of-ecosystems-and-biodiversity-teeb-ecological-and-economic-foundations/* (accessed 5 May 2014).

UNESCO (2012), *Managing Water under Uncertainty and Risk*, The United Nations World Water Development Report 4, Vol. 1, United Nations Educational, Scientific and Cultural Organization, Paris.

Vörösmarty, C.J. et al. (2010), "Global threats to human water security and river biodiversity", *Nature*, Vol. 467, 30 September, *http://dx.doi.org/10.1038/nature09440*.

Vörösmarty, C.J. et al. (2000), "Global Water Resources: Vulnerability from Climate Change and Population Growth", *Science*, Vol. 289, No. 5477.

Water Corporation of Western Australia (2014), *www.watercorporation.com.au/water-supply-and-services/rainfall-and-dams/streamflow/streamflowhistorical*.

Wickel, A.J. and J.H. Matthews (2012), adapted from "Vulnerability to What Change?", presented at the UNFCCC Technical Workshop on Water, Climate Change Impacts and Adaptation Strategies, World Wildlife Fund and Conservation International, Mexico City, Mexico, 18-20 July, *http://unfccc.int/adaptation/workshops_meetings/nairobi_work_programme/items/6955.php* (accessed 12 December 2012).

Winpenny, J. (2011), "Keynote Speech, A Drought in Your Portfolio?", EIRIS Conference, London, 7 June, *www.unesco.org/water/wwap/news/pdf/EIRIS_Conf-JWinpenny_Keynote_Speech.pdf*.

WRI (2015), "Aqueduct: Measuring and Mapping Water Risk", World Resources Institute, *www.wri.org/our-work/project/aqueduct* (accessed 6 January 2015).

Young, M. (2013), *Improving Water Entitlement and Allocation*, background paper for the OECD project on water resources allocation (unpublished).

Chapter 2

A framework for water allocation

This chapter sets out an analytical framework for water allocation regimes as a basis for examining how they function in a range of countries and how they can be improved. It highlights how water is a complex resource, with distinctive features as an economic good, often with a unique legal status. It identifies the key components of an allocation regime and the policy levers that can be used to improve their performance. Finally, the framework links the elements of allocation regimes with the policy objectives of economic efficiency, environmental sustainability, and social equity.

Key messages

- Water allocation is, in essence, a means to manage the risk of shortage and to adjudicate between competing uses via a **combination of policies, laws, and mechanisms**. The risk of shortage is dynamic, in both the short and the long-term. Hence, a well-designed allocation should have two key characteristics: it should be **robust** by performing well under both typical and extreme conditions and demonstrate **adaptive efficiency** with the capacity to adjust to changing conditions at least cost over time.

- Water is a complex resource with **distinctive features as an economic good**, displaying the characteristics of public or private goods in different settings. The public or private good nature of water depends on *how* and *where* water is used or, more precisely, valued (for either use or non-use purposes).

- Water resources usually enjoy a **distinctive legal status**, often managed under the Public Trust doctrine. Access to the resource is often subject to **usage rights** (or "water entitlements"), rather than outright ownership (with the exception of groundwater resources in certain countries).

- **Nested allocation arrangements** can allow for tailoring the design of allocation arrangements to specific settings.

The allocation of water resources determines who is allowed to abstract water, how much can be taken and when, how much must be returned (of what quality), and the conditions associated with use. "Allocation regimes" consist of a collection of public policies, mechanisms, legal and economic arrangements, as well as informal conventional practices. This chapter sets out an analytical framework for allocation regimes that can be used as a basis for examining how allocation regimes currently function in a range of countries and to identify opportunities for improvement. It first highlights the general policy objectives that allocation regimes usually aim to achieve. It then reviews the particular characteristics of water resources as an economic good and the legal arrangements that determine access to the resource. While transboundary allocation is not a focus of this report, some considerations are discussed. Finally, the chapter sets out a framework that links policy objectives to the various components of allocation regimes, to provide a reference against which current approaches can be examined and the levers for improving regimes can be identified.

Policy objectives of allocation regimes

Water allocation is, in essence, a means to manage the risk of shortage and to adjudicate between competing uses. Water systems are inherently variable. The availability of water resources varies in time (seasonal, inter- and intra-decadal variability) and space as a result of the natural water cycle. These features are reinforced and exacerbated by climate change (OECD, 2013a). Freshwater supply conditions can change quickly. It is not uncommon for a region that has been in the midst of a serious drought to find itself suddenly in the midst of a flood.[1]

The risk of shortage is dynamic, in both the short-run and the long-term. Hence, a well-designed allocation regime should have two key characteristics: it should be robust by performing well under both average and extreme conditions and demonstrate adaptivity with the capacity to adjust to changing conditions at least cost over time. Adaptive efficiency addresses the least cost path to maximise social welfare over the long term in the context of complex resources, unpredictability, feedback effects and path dependencies (Marshall, 2005).

Water users need to understand that water supply involves short-term and long-term risks. Neither the average amount of water that can be supplied nor the minimum amount of water that can be supplied from a natural source can be guaranteed. Failure to make the nature of these risks clear can result in over-investment in water-dependent enterprises and calls for compensation when entitlements need to be reduced in order to avoid compromising water quality and other environmental outcomes.

Various water users will have different levels of risk aversion and different risk preferences. For instance, it is critical to avoid any shortage in water availability for cooling nuclear power plants, as the consequences are unacceptably high. However, farmers growing low value annual crops may be willing to forego water use during times of scarcity, especially if they can recover greater value from trading their water entitlements to higher value uses than they can by using the water. Various water users also have different capacities to manage the risk of freshwater shortage, by improving efficiency, relying on alternative water sources, or adjusting the timing of their water use. In principle, to reflect varying risk preferences and capacity, water uses should have flexibility to manage their risk of shortage.

In managing the risk of shortage, water allocation regimes should aim to maximise value individuals and society obtains from water resources in terms of economic, environmental and social outcomes. To achieve this, three generic principles can be used as a guide: economic efficiency, environmental sustainability, and social equity. Well-designed allocation regimes should manage the risk of shortage without compromising the effectiveness of arrangements needed to ensure full control of flood, water quality and other water management tasks.

From the perspective of economic efficiency, two issues should be considered. The first relates to determining an efficient level of water supply that balances the marginal costs of increasing supply or improving its reliability (e.g. through new sources of water, such as wastewater re-use, desalination, inter-basin transfers, or developing storage and distribution infrastructure) with the marginal benefits of doing so. This "supply side" approach has often been a response to increasing pressure on water resources, although it is not always efficient from an economic point of view when greater net benefits can be reaped from improving demand side management. The second issue relates to how existing water resources are allocated among various purposes (including *in situ* uses). Box 2.1 discusses allocative efficiency from a theoretical and practical perspective.

In addition to allocative efficiency, promoting the technical efficiency of resource use is another important dimension. The allocation regime should be designed in a way that encourages efficient water use and promotes innovation to increase the value derived from water use. This requires an allocation regime that provides incentives for efficient resource use and removes perverse incentives for inefficient use.

Appropriate abstraction charges, which reflect at least the full cost of providing access to water resources are an important feature of allocation regimes. In practice, abstraction charges tend to be administrative fees and are not based on the economic value of the water abstracted (Hanemann, 2006). Because water prices have historically been far below efficient levels, increases have the potential to contribute to improving the problem of water shortage (Olmstead, 2010). At the minimum, abstraction charges would provide an incentive to eliminate the most inefficient uses. In addition to allocative efficiency and encouraging efficient resource use, the transaction costs of the overall regime should remain as low as practicable.

Environmental sustainability is a second generic objective of allocation regimes. As pressures on water resources grow, the availability of water to maintain environmental value is increasingly strained. Allocation regimes should have hydrological integrity as well as allocate sufficient water to meet environmental needs, referred to as "environmental flows". The Brisbane Declaration (2007) defines environmental flows as the "*quantity, timing and quality of water flows required to sustain freshwater and estuarine ecosystems and the human livelihood and well-being that depend on these ecosystems*".

In practice, environmental flow requirements can be estimated in a number of different ways, ranging from simple rules of thumbs, to sophisticated, data intensive approaches. Poff et al. (2009) have developed a framework, the ecological limits of hydrologic alteration (ELOHA), for assessing environmental flow needs for many streams and rivers simultaneously to foster development and implementation of environmental flow standards at the regional scale. In order to apply the ELOHA framework, stakeholders and decision-makers need to explicitly evaluate "acceptable risk" as a balance between the

Box 2.1. **Improving allocative efficiency for water resources: A theoretical and practical perspective**

From a theoretical perspective, an allocation would be considered "Pareto optimal" (or "Pareto efficient") in the case where further reallocation among users cannot make anyone better off without making at least one person worse off. A change in the allocation of resources among users (which may include some compensation to offset any loss of welfare of users due to foregoing use of the resource, even if the transfers are not actually made*) that makes at least one person better off, without making at least one person worse off is considered a "potential Pareto improvement". The relative efficiency of alternative allocations can be viewed with respect to this criterion, by determining whether an alternative allocation situation provides a "potential Pareto improvement".

From a practical perspective, achieving this "optimal" allocation for water resources is problematic for numerous reasons. Water is not one single commodity, but instead a complex, multi-dimensional resource (both legally and hydrologically) (Olmstead, 2010). It is also difficult to estimate the marginal benefit (use and non-use values) of water resources in various settings. For example, what is the value of additional water left in-stream to support a freshwater ecosystem? The value obtained from water resources depends on when and where it is used (or not used). In addition, the marginal benefits of some uses (e.g. drinking water, for national security purposes such as flood protection or cooling nuclear plants) can far exceed the financial cost of supplying water for others (e.g. irrigation) (see Briscoe, 1996), complicating attempts to use marginal cost pricing as an allocation mechanism.

While there are practical limitations to achieving "optimal" water resource allocation, improving the efficiency of allocation is an achievable and valuable goal. It is possible to make changes in allocation regimes to make at least one person better off, without making anyone worse off. One way to do this is to limit changes to those that increase the value of the water entitlements held. Another is to offer compensation to those who would become worse off as a result of a change in water allocation. When considering this option, it is important to begin by assessing the *status quo* in the absence of reform, which can provide a baseline to guide action.

* The Kaldor-Hicks compensation principle states that if transfers (of compensation) could be made to achieve unanimity (in terms of social preferences) on a particular choice, then the choice is socially desirable, *even if the transfers are not actually made* (Kolstad, 2000).

Source: Kolstad (2000); Olmstead (2010); Briscoe (1996).

perceived value of the ecological goals and the economic costs involved, taking into account the scientific uncertainties between ecological responses and flow alteration (Poff et al., 2009).

Along with economic efficiency and environmental sustainability considerations, the equity of dimension of an allocation regime is also important. Equity is subject to interpretation in varying contexts. In the context of water resources allocation, it can relate to equity among various groups (e.g. among water users, between current water users and new entrants, among social groups). It can also be understood in terms of outcomes as well as in terms of processes.

Equity in process is important to ensure that existing and potential water users are treated fairly. Equity also requires that changes in allocation arrangements should fully consider the distributional impacts of these changes. However, the use of water allocation to achieve income and other distributional objectives in OECD countries needs careful consideration. In practice, it is an inefficient means of dealing with income inequality and policy objectives related to supporting farmers' livelihoods or achieving food security. There are proven approaches to address affordability issues to ensure that all households have sufficient access to drinking water for their needs. There are also numerous alternatives to support farmers' livelihoods.

Table 2.1 summarises these general policy objectives that can be used to guide the design of allocation regimes. These various (sometimes conflicting or competing) objectives need to be balanced, which requires addressing trade-offs between them. How this is done in a specific context will be strongly influenced by historic circumstances that have shaped existing allocation arrangements, the relative weight given to certain policy objectives over others, and the prevailing political orientation. In practice, water allocation arrangements are often put in place to serve other policy objectives outside of the water domain *per se*, such as food or energy security. For example, countries that are highly dependent on hydropower for their energy supply, such as Brazil, place a high priority on managing water flows and allocation for energy. Serving the needs of other water users is subservient to this objective. These decisions can be informed by economic valuation, but will also be strongly influenced by a process of negotiation among various groups, such as water using sectors, various regions, historic users and new entrants.

Table 2.1. **General policy objectives of water allocation regimes**

Economic efficiency	Environmental sustainability	Social equity
● Allocative efficiency (allocating water to higher value uses). ● Efficiency in water use. ● Efficient allocation of risk of shortage. ● Efficient level of investment in augmenting water supplies and water dependent activities. ● Incentives for innovation and investment. ● Administrative efficiency.	● Hydrological integrity of the system. ● Adequate environmental flows to support ecosystem services. ● Adequate environmental flows to support key freshwater species.	● Equity among water users (or groups of users) and between existing users and new entrants. ● Equity in the process of allocation and re-allocation. ● Fair distribution of costs and benefits of the allocation regime. ● Equitable sharing of risk of shortage. ● Equity between generations (ensuring sustainable use of the resource) or community groups, including Indigenous people. ● Perceived fairness of the allocation regime.

Water: A resource with public and private good characteristics

Water resources exhibit a number of particular characteristics that make them unlike many other economic goods and natural resources. These distinctive features (summarised in Box 2.2) merit consideration in the design of allocation regimes and can explain, in part, the distinctive legal and economic arrangements underpinning them.

The complexity of water as a resource is reflected in its distinctive features as an economic good often with a particular legal status (Hanemann, 2006; Griffin et al., 2013; Barraqué, 2013; Whittington et al., 2013). These features are critical for understanding how the combination of instruments and mechanisms that make up allocation regimes can be designed to meet policy objectives. As argued by Ostrom (2003), the attributes of the goods allocated, along with the rules used for their allocation affect the incentives that users face.

Water resources display both public good and private good characteristics in different settings. Samuelson (1954) made the distinction between private goods, which "*can be parcelled out among different individuals*" and public goods, "*which all enjoy in common in the sense that each individual's consumption of such a good leads to no subtraction from any other*

Box 2.2. **Water's distinctive features**

The proposition that water is an "economic good" is often the subject of heated debate. Between the opposing views that water is an economic good like any other and that water is sacred and cannot be considered a commodity, Hanemann (2006) argues for a position somewhere between these extremes. First, he acknowledges that water is *perceived* to be different from other essential goods (e.g. shelter, food) and the arousal of passionate debate speaks for itself. Second, water has some *economic* features that make is distinctive *as an economic commodity.* Unlike many other economic goods, water is both a private good and a public good. Water also has a number of specific features that bear on its consumption, value, price and so on.

- *The mobility of water:* water flows, evaporates and seeps into the ground. There are also opportunities for sequential use and re-use (although there is often a reduction in the quality or change in location). This makes tracking water flows costly and sometimes difficult. As a result, it is often hard or impractical to enforce excludability (a requirement for private goods).

- *The variability of water:* the supply of and demand for water varies spatially, temporally and often in terms of quality. This generates the challenge of matching supply and demand, which typically requires infrastructure to facilitate storage, inter-basin transfers. It also affects the legal and institutional arrangements for the use of water. For example the intermittent demand for agricultural water is conducive to collective sharing of right of access (as opposed to individual ownership of a property right), as irrigators may, depending on the crops cultivated, draw water from a common resources at different times.

- *The cost of water supply:* water is expensive to transport, relative to its value per unit (as compared to other commodities, such as petroleum or electricity). Water supply is also exceptionally capital-intensive and capital assets used in water supply are an extreme type of fixed, non-malleable, long-lived capital. For many types of water supply and sanitation infrastructure, there are significant economies of scale. There is generally an unusually large gap between short- and long-run marginal costs in water supply. These factors help to explain the significant degree of public provision of water supply.

- *The price of water:* the price paid by most users represents the physical supply costs and not the scarcity value of the resource itself. Water is often under-priced and charges tend to reflect past supply costs, rather than future replacement costs. Although under-pricing is not unique to water, it is a persistent and ubiquitous feature.

- *Essentialness:* water is both an essential final good (as *no* amount of any *other* final good can compensate for having a zero level of consumption) and an essential input to certain production processes (as *no* production is possible when this input is lacking).

- *The heterogeneity of water:* besides just quantity, water's location, timing, quality and variability influence its value. For a given user, one unit of water is not necessarily the same as another unit if it is available in a different location, at a different time, of a different quality or a different probability of availability. As a differentiated commodity, there is no single demand function for water use.

- *The benefits of water use:* there are a number of ways in which a marginal increase in access to water could result in benefits, either directly or indirectly. Yet, the relationship between water and increased productivity is a complex one. In some cases, water may be a necessary, but not a sufficient condition for economic growth. The relationship may even be more complicated, as *multiple* causal pathways between water and growth may exist.

Source: Based on Hanemann (2006).

individual's consumption of that good". This definition focusses on non-rivalry (consumption of the good by one person does not diminish the ability to consume by others) as the distinguishing feature. "Pure" public goods also exhibit non-excludability (the inability to exclude access to the resource). Excludability is mainly determined by the specific characteristics of the resource, the legal framework used to define the ownership of the resource and entitlements to use it, and the mechanisms in place to measure abstraction and enforce the allocation regime. It is basically about the capacity to prevent certain people from using the resource at an acceptable cost.

The public or private good characteristics of water depend on *how* and *where* water is used or, more precisely, valued (for either use or non-use purposes). Water resource use displays varying degrees of rivalry and exclusivity in different settings. The degree of rivalry varies for different types of uses. This mainly concerns the distinction between consumptive versus non-consumptive uses. This is basically about whether one person's enjoyment of the resource diminishes the possibility of enjoyment by others. For instance, drinking water supply to households is rivally consumed relative to others accessing the public water supply in that area. Irrigation use is rivally consumed relative to alternative uses in the area (via evapotranspiration, water embodied in crops), while the remainder is runoff, which is typically altered in terms of quality and location and is not readily available for others to use (Griffin et al., 2013). In such a situation one person's consumption subtracts from the resource available for others to consume.

In-stream uses of water such as navigation, recreation, or the maintenance of environmental flows to support freshwater ecosystems are examples of public good uses of water, which may be non-rivally and non-exclusively enjoyed (Griffin et al., 2013; Hanemann, 2006).[2] In other words, these types of uses are freely accessible to all and their enjoyment by one user does not impinge on the enjoyment by other users. However, even these uses tend to be non-rival only within their use category, as in-stream use is rival with the out-of stream uses to which this water could be alternatively applied in its locale (Griffin et al., 2013). Table 2.2 summarises these examples of various uses of water resources.

Table 2.2. **Water as a public and private good**

	Rivalry		
	Low	–	High
Excludability **High** – **Low**	**Club good** ● Recreational use in a water bodies where access can be restricted, such as a private lake.		**Private good** ● Water body exclusively on private land. ● Drinking water consumed by households. ● Irrigation system which allows for exclusion. ● Rainwater captured on private land.
	Pure public good ● In-stream uses, such as: navigation, environmental flows supporting ecosystem services, recreational uses in a public setting, such as bathing, boating, etc.		**Common pool resources** ● Shared aquifer. ● Water provided through a distribution system in an irrigation district (where users cannot be excluded).

Source: Based on Griffin et al. (2013); Hanemann (2006).

Excludability, in the case of water resources, is mainly determined by the specific characteristics of the resource itself and the legal and policy framework in place to limit access to it. For example, a water body fully contained on private land with well-enforced property rights exhibits excludability, because other potential users can be preventing from using it. However, a groundwater aquifer that can be freely accessed by multiple users does not. This is an example of a common pool resource; characterised by high exclusion costs and where one person's consumption subtracts from the total available (Ostrom, 2003).

The distinction between public and private characteristics of various uses of water resources is an important consideration for allocation for two key reasons. Firstly, the valuation of public and private goods is fundamentally different: for private goods, the marginal value will be that of a single user (allocated efficiently, it will be the user with the highest and best use), while the marginal value placed on a public good is that of many people who can enjoy the good simultaneously. This basic distinction regarding valuation is one of the reasons why the non-market benefits of environmental protection can outweigh the use benefits associated with diverted use (Hanemann, 2006). Olmstead (2010) reviewed the substantial literature on the marginal value of surface water left in-stream for recreation, wetlands restoration, and other in-stream activities. For example, in the United States, the marginal value for fishing was found to be greater than the marginal value for irrigation in 51 of 67 river basins with significant irrigation, but values were highest in the Southwest, where the effects on fishing of marginal changes in stream flow were the greatest (Olmstead, 2010).

Secondly, since water resources display both private and public good characteristics in different settings, nested allocation arrangements can be appropriate to tailor the design of allocation arrangements to specific settings. For example, administrative allocation can be used to determine the repartition between *in situ* and diverted uses. Once this repartition has been made, market-based allocation can be used where appropriate to allocate water among a specific user group, such as irrigators. This reflects what often occurs in practice. The allocation between environmental flow requirements (along with other *in situ* uses) and water available for diverted uses (irrigation, industry, domestic use, etc.) nearly always occurred administratively, via regulation. In practice, most (but not all) market-based allocation arrangements allow for trade among a specific user group, such as irrigators.

Legal status of water and claims to use water

The characterisation of water as an economic good and the ownership institutions of water resources are two distinct matters. Public ownership of water does not mean that water is necessarily a public good (Griffin et al., 2013). In the same vein, common-pool resources are not necessarily associated with common-property regime, *or with any other particular type of property regime* (Ostrom, 2003). However, the public good nature of water *in situ* has had a decisive influence on the legal status of water (Hanemann, 2006).

Water law has been influenced by a number of legal traditions over the centuries that still have repercussions for its use today. Common-law traditions developed a body of legal doctrine specifying strong property rights in flowing water attached to riparian possession and limited rights to surface and underground waters. Water doctrine was also influenced by Roman law and Roman-derived civil-law concepts of common goods and the natural rights of ownership (Getzler, 2004). Customary rights to water use, with roots in traditional land tenure practices, are typically based on long standing non-state law, custom and traditions (Le Quesne et al., 2007). As documented by Ostrom (1992), a typical case of customary rights is that of communal rights, in which water is allocated by a community via community leaders or a bargaining process to individual users. Table 2.3 summarises different generic types of ownership systems, with an example of application to water

Table 2.3. **Types of property ownership systems**

Type of ownership system	General description	Example of application to water resources
Open access	No defined users or owners. There is no incentive for any one user to protect the resource unless all the users protect it.	Instances where no informal or formal control on access to water resources exists.
Common property[1]	A management group has the right to exclude non-member. Non-members have a duty to abide by the exclusion. Co-owners comprise the management group and have rights and duties with respect to the use of resources.	Collectively owned ground or surface water resources, with entitlements to use the water proscribed to individuals by the management group responsible. Communal rights.
State property	Individuals have a duty to observe the rules of resource use determined by the controlling agency. For state ownership to work efficiently, the state must be able to monitor the use of resources, establish rules acceptable to individuals and communities, and enforce those rules.	Publicly owned ground or surface water resources, with entitlements to use the water proscribed by the public authority responsible.
Private property	Individuals have the right to undertake "socially acceptable" uses, and the duty to refrain from "unacceptable" uses. Others have the duty to respect individual rights. This is likely to conserve the resource since the owner would be able to receive the benefits of conservation. However, markets will still be unable to account for externalities.	Privately owned groundwater resources, or surface water resources contained on private land.

1. Getzler (2004) asserts: "The communitarian theory of the self-regulating commons is not in tension with the conventional economic theory of property but is a special case. The conventional theory postulates that parties move out of common pooling into privatised property regimes in situations where trust and stinting cannot be generated between self-interested parties (see Libecap, 1989). The commons theory locates instances where self-regulated pools become possible through the constraints of rational self-interest buttressed by culture or repeat transacting; privatisation is then unnecessary or even sub-optimal."

Source: Based on Pearce and Özdemiroglu (1997) for the categories of ownership systems and general descriptions.

resources. This is not a comprehensive list of ownership and management arrangements for water resources. Cases also exist whereby the ownership of water resource is not explicitly defined, but the government manages the resource and usage rights.

Flowing waters are often treated as common to all, *res communis omnium*, and are not capable of being owned (Hanemann, 2006) or as *res nullius* ("ownerless property"). Water is thus often the subject of *usufructuary rights*, which allow for the right of use of a resource or property and the enjoyment of benefits from that use. These rights of use are typically subject to a number of conditions, including "reasonable" or "beneficial" use doctrine and limited to a pre-determined duration. The reasonable use doctrine allows an upstream user to interfere with the stream to a certain (vaguely defined) extent, without requiring bargains to be struck with all affected users downstream (Getzler, 2004).

There are several ways in which these rights of use can be defined. In this report, we refer to them as "water entitlements".[3] The conditions applied to them can vary considerably from case to case. Common instances include *riparian entitlements*, under which all landowners whose property adjoins a body of water have the right to make reasonable use of it and often requires that the use does not interfere with other riparians' use of the resource. Water entitlements based on *prior appropriation* results in a continuum of senior right holders to junior rights holders. Appropriative entitlements are assigned in order of application of a quantity of water for a beneficial use. Those applications submitted earlier will be more senior to those submitted later ("first in time, first in line"). Water is then allocated according to seniority. In an extreme drought, even "senior" entitlement holders may not receive their allocation. In a mild drought, all but the most junior entitlement holders may receive full allocations. This system means that more junior users bear a greater risk of water shortage, while more senior users are relatively more insulated from such risk.

Water entitlements can also be *"unbundled"*, or separated, from land titles. Water users hold an entitlement to abstract and use water from a specified resource pool, independently of ownership of land. The entitlement holder is then given a specific "allocation" of water to abstract within a specified time period in a manner that is in accordance with pre-specified conditions. The separation of water entitlements from specific allocations within a season provides a mechanism to combine flexibility within the system to adjust to changing conditions (by varying the allocations within a given season according to the available resource pool) with security for water entitlement holders (by maintaining the overall entitlement over time).

In many instances, several laws may be in place, which generate contradictory or competing claims on water resources. For example, statutory water law and customary water law may co-exist, with varying levels of actual implementation on the ground. This "pluralism" of water law can be further exacerbated by water laws in contradiction with land laws, or environmental laws, resulting in unclear and competing water law systems in a given context and expected management problems (Le Quesne et al., 2007).

Transboundary considerations[4]

In transboundary river basins, each riparian State is entitled (and limited) to an equitable and reasonable share of beneficial uses of transboundary waters (Wouters et al., 2005). The key principles of international water law – equitable and reasonable use of transboundary water and the obligation not to cause significant harm – provide guidance for allocating water, as well as for the development and use of river basins. There are no universally accepted criteria for allocating transboundary waters or their benefits. Each transboundary basin has its own particularities. Arrangements or agreements reached with regard to certain basins cannot be automatically applied to other transboudary basins. The principles of international water law need to be interpreted in the context of each basin's unique setting.

Allocating water between riparian countries can be highly challenging, especially if the water management practices as well as the main water needs in the riparian countries differ, for example with regard to seasonal discharges as it is commonly the case for water use for agriculture and hydropower. A broader understanding of the benefits of water use can lead to more creative arrangements. Some agreements acknowledge existing uses and some specify a minimum cross-border flow to downstream riparians to ensure the availability of a certain proportion of water resources for the downstream riparian countries. Some agreements also specify the obligation to maintain a minimum environmental flow, such as in the Agreement between the Government of the Russian Federation and the Government of Mongolia on the Protection and Use of Transboundary Waters of 1995.

International guidelines and conventions on freshwater set out principles and rules, which can be helpful in guiding decisions about the use of transboundary waters. The Helsinki Rules on the Uses of the Waters of International Rivers,[5] adopted by the International Law Association in 1966, is a pioneering example of such a guideline. Although the guideline is not enforceable, it preceded and informed the establishment of

   43

international conventions on freshwater, namely the Convention on the Protection and Use of Transboundary Watercourses and International Lakes (Helsinki Convention, 1992) and the Convention on the Law of the Non-Navigational Uses of International Watercourses (UN Watercourses Convention, 1997). Although not all OECD member countries are parties to them, some of the provisions of these conventions have provided the basis for a number of bilateral and multilateral agreements on transboundary waters.

In negotiating agreements on water allocation, hydrographical arguments (shares of the States' territories contributing to flow formation) and the securing of existing uses or needs (irrigable land, population or the requirements of a specific project) are commonly referred to (see e.g. Wolf, 1999). Agreements with clauses for water allocation have commonly been negotiated in conjunction with a boundary delineation, a division of boundary waters or an agreement over future river development. In response to declining groundwater tables resulting from heavy abstraction, an arrangement was concluded in 1977 between France and Switzerland on the Genevese aquifer defining an abstraction cap and a level up to which abstraction is free of charge.[6] Under the Arrangement, the Swiss constructed and have operated an artificial recharge installation and the costs of construction and operation were agreed to be shared by the parties (Mechlem, 2012). This illustration shows how investments for mutual benefit may facilitate agreement on water allocation.

Institutional arrangements at a transboundary level that allow for flexible responses to changing circumstances can be useful, such as through continued co-operation in technical subsidiary bodies (UNECE, 2009). As the availability of water resources varies over time due to climate change, shifting hydrological conditions, land use and land cover change, as well as development, agreements on transboundary waters may also need to adjust in response to these changes in order to effectively manage risk (see example in Box 2.3). Because such agreements are difficult to renegotiate and tend to last for decades, it is better to define water allocation at the transboundary level in a flexible manner considering supply and demand balance.

Policy instruments and mechanisms for water allocation

The complex and distinctive features of water resources as an economic good and its particular legal status mean that allocation regimes are often complex combinations of various laws, policies, and mechanisms. This section identifies the main policy instruments and mechanisms used to allocate water and describes their constituent parts. The robustness and adaptive efficiency of an allocation regime can be improved by unbundling the various elements and using separate instruments to pursue various objectives (Box 2.4). However, unbundling should not undermine the effective management of the system as a whole. Thus, even if separate instruments are used to manage particular objectives, there is still a need for a comprehensive view of how the various elements interact to achieve policy objectives.

The elements of an allocation regime can be divided into "system level" elements and "user level" elements (Young, 2013). System level elements include those issues that are most efficiently and equitably dealt with at the scale of a water resource (be it at the basin, catchment, river, stream or aquifer level). Typically they take the form of conditions expressed in water law, water sharing plans and other similar policy instruments that determine how system wide decisions are taken and by whom. Those that apply to all water resources may be set out in regional and national legislation. System level

> **Box 2.3. Improving responsiveness to variability and hydrological extremes in transboundary allocation: The Albufeira Convention**
>
> In river basins where water is scarce or its availability variable, defining water allocations between the riparian countries in terms of an annual allocation may not be sufficient for managing variability. The distribution between seasons is important for several types of water use, including agriculture and hydropower. The Albufeira Convention provides a good example of shifting from allocations based on annual flow to shorter time periods, to better enable adjustment to variability and extremes.
>
> The Convention on Co-operation for the Protection and Sustainable Use of the Waters of Portuguese-Spanish River Basins (Albufeira Convention, 1998)* was revised in 2008 to guarantee a good status of water bodies and to meet the current and future demands, including responding to floods and droughts. The Albufeira Convention regulates the transboundary waters in the shared basins between Spain and Portugal: the Miño/Minho, Lima/Limia, Douro, Tejo/Tajo and Guadiana Rivers.
>
> The original Convention defined the absolute amount of water that should be received by the downstream riparian State. The annual flow regime defined in the Convention operated well for many years and good collaboration helped the riparian countries to overcome difficult situations. Nevertheless, the exceptional drought period that affected the Douro, Tejo and Guadiana basins from 2004 to 2005 demonstrated the potential impacts of water scarcity. The amendment of 2008 involved disaggregating the annual flow regime into shorter time periods. This new regime determines a quarterly (Minho, Douro, and Guadiana), weekly (Douro and Tejo) and daily (Guadiana) discharge flow, depending on the rainfall conditions in each basin (Otterman and Koeppel, 2014). At times of extremely low rainfall (below specific thresholds), the defined flow regime might not apply, but water should be managed in such a way as to ensure its priority uses.
>
> Operating and controlling such a jointly-defined flow regime requires good monitoring capacities and infrastructure, as well as flow regulation, to be in place. Some of the Spanish-Portuguese rivers have significant storage capacity, notably the Guadiana River (UNECE, 2011), so the natural discharge can be complemented with releases from the Spanish reservoirs. Well-functioning information exchange is a prerequisite for an effective, co-ordinated response to hydrological extremes that brings mutual benefits.
>
> * "Convenio sobre cooperación para la protección y el aprovechamiento sostenible de las aguas de las cuencas hidrográficas hispano-portuguesas", Albufeira, 30 November 1998. Available at: *www.boe.es/boe/dias/2000/02/ 12/pdfs/A06703-06712.pdf*.
>
> *Source:* Otterman and Koeppel (2014); UNECE (2011).

elements range from identifying the availability of water resources, to the legal status of the resource, to mechanisms for monitoring and enforcement. Table 2.4 provides a summary of the various system level elements of allocation regimes and provides a brief description of each. Boxes 2.5 and 2.6 provide specific examples of how certain countries deal with two of these system level elements, the assessment of water resources and dealing with in-stream flows.

User level elements of a water allocation regime are those aspects that are most efficiently and equitably dealt with by specifying the arrangements that apply to an individual (or collective) abstractor. Typically, these take the form of arrangements specified in entitlements, permits and licenses. These are summarised in Table 2.5.

> ### Box 2.4. **Advantages of "unbundling" to encourage innovation and investment**
>
> Throughout much of the world, entitlements to abstract water are typically bundled together with controls on use and other arrangements, which can decrease flexibility and can discourage innovation. For example, time-bound duration of water entitlements (often 5-10 years) can reduce incentives for long term investment and innovation of water-intensive activities. Water entitlements are often granted with a limited duration to allow for revision of the conditions on the entitlement and an opportunity to deny renewal if needed. However, from the perspective of the water user, this can generate uncertainty. One option to address this is to issue entitlements in perpetuity, yet ensure that they remain subject to specific conditions. This can only be achieved if entitlements are unbundled from other elements of the allocation regime.
>
> Unbundling involves the replacement of a single licensing instrument with a suite of instruments, each designed to address a specific issue, at the appropriate scale. The can bring considerable advantages. Consistent with the Tinbergen Rule that stipulates that for each and every policy target there needs to be at least one policy instrument, separate instruments for allocation should be used for independent objectives. For example:
>
> - Water sharing plans can be used to define abstraction limits for each water body.
> - Water entitlements can be used to specify each user's priority share in a water resource as well as those people who have a direct interest (or claim) in that entitlement.
> - Allocations can be used to determine the amount of water that an entitlement holder can be granted permission to abstract within a specified time period in a manner that is in accordance with pre-specified conditions.
> - A set of "approvals" (such as abstraction approvals, works approvals and land use approvals), or instruments to specify conditions of use, can be used to monitor water take, control local impacts, and adjust for return flows.
>
> The result is a combination of arrangements that enables clear signals to be given to all water users without having to be addressed by one centralised office. For example, basin planning can occur at the basin level and using well-defined sharing arrangements and catchment scale arrangements can then be nested inside a basin plan that allows for changes at the basin scale to flow through the system, without the need to adjust the catchment plan. Water entitlements can then be defined for each water body so that allocations in a given season can be made in a way that is consistent with these plans. Then, local administrators can be left to manage the local impacts of water use without the need to worry about broader catchment and basin-wide issues, improving the efficiency and effectiveness of governance arrangements.
>
> In Australia, the unbundling of abstraction licensing arrangements has increased the flexibility of the system and caused a significant increase in the value of water entitlements. For most of the last decade, the internal rate of return from water entitlements remained above 15% per annum (Bjornlund and Rossiter, 2007).
>
> *Source:* Young (2013).

Table 2.4. **Description of key system level elements of a water allocation regime**

System level elements of an allocation regime	Description
Legal status of the ownership of water resources	Legal definition of the ownership of water resources (e.g. public, private, *res nullius*).
Institutional arrangements for allocation	Authorities and organisations responsible for allocation and their various roles (policy, planning, issuing entitlements, monitoring and enforcement).
Identification of available water resources	Identification of available water resources (surface, ground water as well as treated waste water intended for re-use) based on best available scientific evidence.
Identification of in situ flow requirements/Identification of available ("allocable") resource pool	An explicit definition of *in situ* flow requirements based on various factors, such as requirements for base flow, environmental flows, non-consumptive use, international commitments, inter-annual and intra-annual variability, and climate change. The remaining water would be considered the available resource pool.
Abstraction limit ("cap")	An explicit and enforceable limit on abstraction. It may be defined in absolute, volumetric terms or as a proportion of available resources. The "cap" can be used to ensure water for environmental needs, so it should be designed to reflect natural flow regime dynamics.
Definition of permitted uses not required to hold an entitlement	Definition of those water users and uses that are allowed to access and use water without holding an entitlement.
Definition of "exceptional circumstances"	An explicit definition of circumstances that are considered "exceptional" and may require extraordinary measures. Stakeholders may or may not be involved in the definition of what constitutes "exceptional circumstances".
Sequence of priority uses	A pre-defined sequence of priority uses sets out the priority of access to water according to types of uses or users. It may apply when "exceptional circumstances" are declared or be used to guide the allocation of water entitlements.
Requirements for new entrants or expanded water entitlements	Conditions placed on the acquisition of new water entitlements or requests to expand existing entitlements. Typical examples include the assessment of third party impacts, environmental impact assessments or existing users foregoing use (for instance, in situations where the catchment is closed).
Mechanisms for monitoring and enforcement	Mechanisms such as metering, aerial surveillance or other means of monitoring water abstraction and use as well as clearly defined procedures and sanctions for addressing infractions and resolving conflicts.
Appropriate infrastructures	Water infrastructures to allow water to be stored, treated and transported, as needed.

Box 2.5. **Assessment of water resources: Examples from Spain and France**

In Spain, the evaluation of available resources and demands in each water resources system are carried out in the River Basin Management Plans (RBMPs). An inventory of available water resources is produced and existing water uses and demands are identified. Water resources assessment methods have been developed for the whole Spanish territory, as well as simulation water resources models that take into account: conventional and non-conventional water resources, environmental flows, water demands, hydraulic infrastructure, water use priorities and exploitation rules in order to establish water allocations and reserves.

In France, water is generally abundant, although water stress is increasing in some regions and there are periodic episodes of scarcity. In areas of water stress, more detailed assessments of water availability and use are justified. Laws related to water management of water resources are stricter in these areas and the allocation regime is more rigorous.

A mapping exercise has been undertaken to identify ground and surface water stressed areas. This is used to define water apportionment areas, where the water deficit is structural. These zones are the target of recent reforms that aim to restore sustainable abstractable volumes as well as the creation of Single Collective Management Bodies (Organismes Uniques de Gestion Collective, OUGC) to provide an incentive for irrigators to allocate a set volume of water among themselves at catchment level.

> ### Box 2.6. **Options for treatment of in-stream flows within a water allocation regime**
>
> When designing an allocation regime and setting a long-term abstraction limit, it is important to decide whether or not to include some or all entitlements in this limit. The most common approach is to set aside the amount needed for environmental needs, non-consumptive uses, and transfers to other systems (including downstream obligations) as a prior right and then to allocate the remainder to take water for consumptive purposes.
>
> An alternative approach, being tested in Australia, is to assign some water to the environment as an entitlement to a share of all inflows and define this entitlement separately from the arrangements used to ensure that base flows, for example, are maintained. In the Murray-Darling Basin, a Commonwealth Environmental Water Holder has been established and by 2019 is expected to hold around one third of the Basin's water entitlements. Under this new arrangement, it is not possible for the government to allocate water to consumptive users without making a *pro rata* allocation to the Commonwealth Environmental Water Holder.
>
> Australia is moving to this approach in order to put environmental water on the same footing as all other water users. Under this arrangement, allocations are made in proportion to the number of entitlements held in the interests of the environment, no matter how dry or wet it is. As a result, administrators are not able to transfer environmental water to other users.
>
> In the United States, non-governmental groups have been buying water to ensure that the environment is looked after. A well-known example is the Oregon Water Trust, which became a programme of The Freshwater Trust in 2008.
>
> *Source:* Young (2013).

Table 2.5. **Description of key user level elements of a water allocation regime**

User level elements of an allocation regime	Description
Legal definition of water entitlements	A legal definition of water entitlements that confers the right to use the resource, usually under certain conditions as well as identification of the types of water users that are required to hold an entitlement in order to abstract water. Entitlements may or may not be privately held. They can be granted to individuals or to collective bodies, such as water users' associations, which then distribute water to individual users. The definition usually also determines how an entitlement can be withheld or cancelled.
Abstraction charges	Charges associated with water abstraction. They aim to recover costs and to internalise negative externalities associated with water abstractions. As a proxy, most charges are set administratively and are designed to recover the costs associated with water supply provision.
Specification obligations relating to return flows and discharges in water entitlements	Return flow obligations refer to the requirement to return a portion of the water abstracted to the same or a different water body following use. Discharge requirements relate to the quality requirements (including thermal changes) of discharges.
Duration for water entitlements, accompanied by expectations for renewal	The length of time a water entitlement is granted for. It may be for a given number of years or in perpetuity (often conditional on beneficial use).
Possibility to trade, lease or transfer under appropriate conditions	The ability for water entitlement holders to trade (either permanently or temporarily), lease or transfer entitlements to others.

Among the user level elements, addressing return flows can be particularly challenging. When water becomes scarce, entitlement holders have an incentive to reduce return flows and save the water for themselves. This can undermine the integrity of the allocation regime if the change in the effective rate of consumption is not accounted for. There are generally two approaches to address this issue: i) reducing the abstraction limit as the technical efficiency of water use increases, with the reduction averaged across all entitlement holders equally; or ii) specifying return flow obligations in water entitlements.

Choosing between these options depends on an assessment of administrative costs and preference for stimulating innovation. The first approach rewards first movers in the pursuit of technically more efficient uses of water. The rate of adoption of more efficient irrigation technology should be faster. Those that move first, benefit from access to water that was previously being used by others. The latter approach is more equitable, as changes in the choice of technology made by one person, which increase the technical efficiency of water use, have no impact on the amount of water allocated to all other users, as would the case in the first approach. However, the latter approach is much more expensive to administer as the type of technology used by each person needs to be tracked and accounted for. In some cases, including several parts of the United States, a hybrid approach is taken. No attempt is made to account for changes within a farm, but when an entitlement is transferred to another person the entitlement is adjusted for expected changes in the return flow. Box 2.7 provides an example from South Australia.

Box 2.7. **A pragmatic approach to dealing with return flows: An example from South Australia**

The groundwater entitlement regime used in the South East of South Australia provides a good illustration of mixing pragmatism with the best available science to account for return flows. The regime began simply and has evolved with a considerable degree of pragmatism. In the first instance, each irrigator was given an entitlement to the volume of water needed to grow a standard crop and this was done by issuing irrigation equivalents. This was seen to be fair and no person would be able to change crop type or irrigation practice in a manner that undermined the interests of others.

Each irrigation equivalent entitled its holder to take the water needed to grow one hectare of a "standard crop" efficiently. Rather than controlling land-use tightly, the volume of water needed to grow a standard crop was estimated and a set of look-up tables used to specify how many hectares of any crop this would entitle its holder to irrigate. The focus of this approach was on the net amount of water used. That is, there would be no penalty for using irrigation techniques that resulted in the return of a significant amount of water to an aquifer.

Just over a decade ago, it was decided it was time to transition from this simple area-based abstraction regime to a more elaborate volumetric regime. This change was motivated by the impetus to reward more efficient water use, reap the benefits of fixing over-allocation problems and allow land-use change.

Box 2.7. **A pragmatic approach to dealing with return flows: An example from South Australia** (*cont.*)

Conversion from an area-based to a volumetric abstraction regime took around four years. At the start of this period, all irrigators were required to install meters and provide information on the amount of water they were using. Local researchers collected information on crop water use from the United Nation's Food and Agriculture Organisation (FAO) and supplemented this with data collected from field trials, the best available science on return flows, and data obtained from local farmers. Considerable care was taken to involve irrigators in all aspects of the conversion programme.

During this process, care was taken to ensure that changes in irrigation practice and special crop water needs were fully accounted for. The model developed for volumetric conversion provided a delivery component for drip irrigation of 11% over and above the crop water use. The delivery component for spray irrigation was set at an extra 18%. There was also a delivery supplement for flood irrigation that depended upon soil type which ranged between 54% and 199%. All water used for drip and spray irrigation was assumed to be lost from the aquifer. In the case of flood irrigation, the volume of water in excess of that required to carry out spray irrigation was assumed to return to the source aquifer and was therefore not accounted for as a net loss to the resource.

Towards the end of the conversion programme, it became clear that the level of allocation following volumetric conversion would be unsustainable if fully extracted. As a result, all allocations were then reduced on a *pro rata* basis to bring use within a sustainable level of extraction, in line with Australia's National Water Initiative.

Allocations are now issued on the basis that accounts for net losses from irrigation plus a delivery component. Allocations representing the net amount of water used are tradeable. Delivery volumes associated with returns to the aquifer are not tradeable.

During consultation, in the district of Padthaway, the community developed an approach based on differential reductions in order to ensure final allocation was within sustainable limits (almost 50% reduction in allocation) while still preserving the economic output of the region. All licensees received a minimum tradeable allocation of 3.95 ML/irrigation equivalent, reflecting the almost 50% reduction required (sufficient to drip irrigate vines). Active irrigators received additional volumes based on their history of use in the last three years, therefore protecting the economic output of the region. An upper cap on allocation was set at the maximum volume that would have been issued through the volumetric conversion model or the maximum volume used by the irrigator, whichever was lower (therefore, more efficient irrigators received less, as they did not need the additional volume). Temporary (two years) bridging allocations were available to inefficient irrigators to catch-up with best practice, etc.

Finally, the current plan requires that the numerical model used to assess return flows and net effects on use be re-run every 5 years with updated data. Any further changes required in order to keep use within sustainable limits will apply to all irrigators proportionately. A re-run of the numerical model in 2014 has indicated that no further reductions are required at this time.

Source: Based on DWLBC (2006a); DWLBC (2006b); South East Natural Resources Management Board (2011).

In addition to defining returns flow requirements from a quantitative perspective, the quality requirements (including thermal changes) for discharges need to be addressing. Box 2.8 provides an example from Spain.

> **Box 2.8. Water discharge permits: An example from Spain**
>
> In Spain, as a general rule, the granting of a water use licence does not imply the granting of a discharge permit for the water after it has been used and the quality has deteriorated. This wastewater must be returned to the natural environment (rivers or aquifers) under quality conditions prescribed by an administrative authorisation via a discharge permit. These discharge permits are granted for a much shorter period than water use licences. They have to be compatible with the receiving environment and are regulated by a list of "limit concentration values" for the main physical-chemical parameters. They are registered in a "Wastewater Census" and are subject to the payment of a Wastewater Control Fee.
>
> In the case of the main surface water resources, the withdrawal volumes are fixed by the Withdrawal Commission (a management body of the River Basin Authority) according to the users' entitlements and the resource availability for a specific period of time. The Withdrawal Commission is in charge of setting the filling and discharge regime governing the reservoirs and aquifers of the river basin in order to comply with the licensing entitlements of the different users.

To summarise how the various elements of an allocation regime fit together, Figures 2.1 and 2.2 provide an overview of the system level and user level elements, respectively. As indicated above, while unbundling can be used as an option to use separate instruments to pursue various objectives, there is still a need for a comprehensive view of the system as a whole, to ensure that the interaction of the various elements works together to reach overarching objectives.

To understand how these various elements of an allocation regime can either help meet or impede policy objectives, Table 2.6 brings these elements together with the policy aims of economic efficiency, environmental sustainability, and social equity. The presence of a given element within an allocation regime does not necessarily ensure that the specific policy objective will be achieved. However, it is the design of the element that will influence its contribution to stated objectives. The descriptions provided in Table 2.6

Figure 2.1. **System level elements of a water allocation regime**

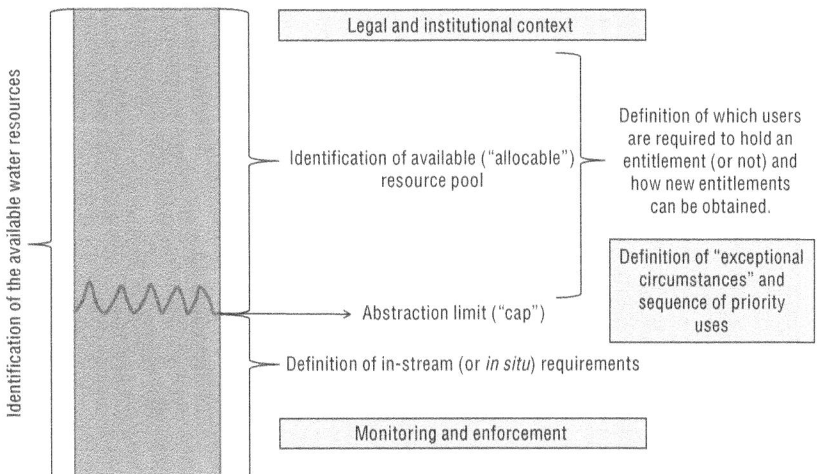

Figure 2.2. **User level elements of a water allocation regime**

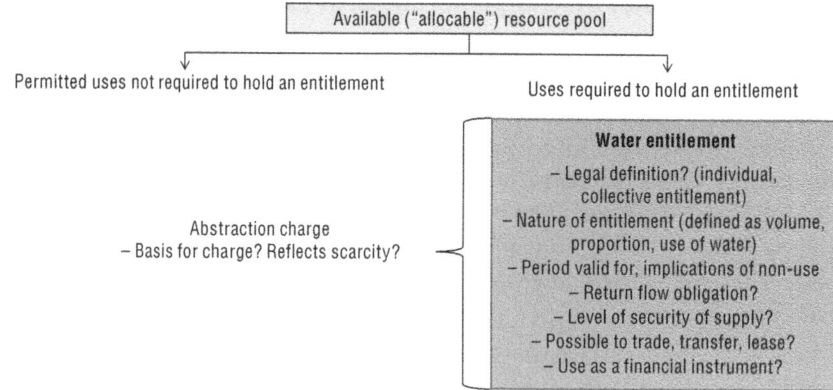

Table 2.6. **Framework for water allocation regimes**

Elements of an allocation regime	Economic efficiency	Environmental sustainability	Social equity
	System level elements		
Legal definition of the ownership of water resources	Allows for clear assignment of entitlements of use.	Confers legal authority to secure water for the public good uses.	Allows for clear assignment of entitlements of use.
Appropriate institutional arrangements for allocation	Ensure that a competent public authority can manage system and user level allocations issues, with clear lines of accountability, while minimising transaction costs.	Ensure that a competent public authority can designate and enforce adequate environmental flows.	Ensure equity in process, through adequate mechanisms for stakeholder engagement.
Identification of available water resources	Allows for efficient augmentation of available resources.	Ensures hydrological integrity and allows for managing system connectivity.	May be used to ensure fair access to adequate resources.
Identification of in situ requirements/definition of available ("allocable") resource pool	Balances use and non-use values of *in situ* activities vs. use value of diverted activities.	Ensures adequate environmental flows.	Balances needs of *in situ* users vs. users of diverted flows.
Abstraction limit ("cap")	Balances cost of closing system with risks of unsustainable use.	Allows for "closing" the resource pool to ensure sustainable functioning.	Balances needs of current water users with future water users.
Definition of permitted uses not required to hold an entitlement	Balances transaction costs associated with managing small scale uses with the cost (risk) of possibly undermining system integrity and foregoing abstraction charges.	Ensures hydrological integrity.	Balances customary, small-scale and subsistence uses with need with the need for system level integrity.
Definition of "exceptional circumstances"/sequence of priority uses	Can be used to ensure that the sequence of priority uses reflects to some extent the marginal value of use.	Can be used to avoiding irreversible damage to vulnerable ecosystems and key species and ensure environmental flows are not simply used as an "adjustment factor" in times of scarcity.	Can be used to ensures that human needs are a priority; Equity in process can be ensured through involvement of stakeholders in the definition of "exceptional circumstances" and the sequence of priority uses.
Requirements for new entrants	Ensures that water can be allocated to higher value uses.	May require an environmental impact assessment to support hydrological integrity.	May require an assessment of third party impacts; can be used to encourage fairness in access between existing users and possible new entrants.
Mechanisms for monitoring and enforcement	Balance transaction costs associated with monitoring and enforcement with costs (risks) imposed by unauthorised use.	Ensure hydrological integrity by ensuring adequate environmental flows.	Ensure common pool resources are used equitably and that use entitlements are respected (amounts are not exceeded) and illegal use discouraged.
Appropriate infrastructures	Ensure that water can be stored, treated and transported to water users as needed.	Ensure that water to serve environmental purposes can be stored, treated and transported to water users as needed.	Ensure that water uses have adequate and fair access to water.

Table 2.6. **Framework for water allocation regimes** (*cont.*)

Elements of an allocation regime	Economic efficiency	Environmental sustainability	Social equity
User level elements			
Legal definition of water entitlements	Provides incentives for investment and innovation.	When entitlement defined as a proportion of available water, contributes to hydrological integrity and allows for adjustment to changing conditions.	May contribute to equity in processes that determine the conditions applied to water entitlements.
Abstraction charges	Promotes recovery of costs associated with providing freshwater supply, along with the environmental costs of resource use and (possibly) scarcity value.	May be used to reflect the environmental costs of resource use and (possibly) scarcity value in charges appropriate for hydrological conditions.	May be reviewed for potential affordability issues.
Specification of return flows obligations in water entitlements (quantity and quality)	Provides incentives for efficient water use.	Contributes to hydrological integrity and managing the quality of discharges.	Allows for a positive externality to be reaped by more efficient users.
Appropriate duration for water entitlements with clear expectations and transparent process for renewal	Provides an incentive for investment.	Contributes to hydrological integrity.	Contributes to equity in process related to conditions renewals.
Possibility to trade, lease or transfer under appropriate conditions	Encourages allocative efficiency, provide an incentive for efficient water use and innovation.	Allows for water "buy-backs" to occur, which can secure water from existing waters to be reallocated for environmental purposes thereby Increasing flexibility in managing scarcity conditions.	Allows for more flexibility in sharing of risk of shortage.

illustrate how the various elements of an allocation regime can support policy objectives, if they are well-designed. The "health check" set out in Chapter 5 provides policy guidance on how the design of allocation regimes could be improved.

Conclusion

Water is a complex resource, with distinctive features as an economic good and often with a unique legal status. Water resources display varying degrees of rivalry and exclusivity, depending on the type of use (or non-use). Allocation regimes need to consider both public and private good characteristics of water resources since these will bear on how the use (or non-use) of the resource is valued in a particular setting and who reaps the benefits. Nested allocation arrangements can allow for the tailoring of the design of allocation arrangements to specific settings.

Allocation regimes consist of various elements (policies, laws, mechanisms), that, when well-designed, can help to achieve the policy objectives of economic efficiency, environmental sustainability, and social equity. These various and sometimes conflicting or competing objectives need to be balanced, which requires addressing trade-offs between them. The analytical framework set out in this chapter identifies the various elements of an allocation regime and makes an explicit link to the policy objectives that they may influence. This provides a reference against which current regimes can be examined and the levers that are available to improve them. The following chapter will examine the current allocation landscape, as represented by 37 examples of allocation regimes in 27 countries, according to the elements of this framework.

Finally, given the inherent variability of water resources and shifting pressures and social preferences, water allocation regimes also need to be both robust and demonstrate adaptive efficiency. This requires striking a balance between the need for flexibility at the system level and security at the user level, giving both water managers and water users greater capacity to manage risk. Chapter 5 sets out the *"Health Check" for Water Resources Allocation* based on this framework that can support policy makers in their efforts to improve the performance of allocation regimes.

Notes

1. The case of Brisbane, Australia is instructive, as the city has experienced significant problems with both drought and flooding over the past 40 years. The Wivenhoe dam was built to mitigate risks from both floods and drought. In 2008, during a period of prolonged drought, the water level fell to around 17% capacity, leading to a focus on managing water scarcity. However, several months of intense rain in 2010 increased the water level so rapidly, that it resulted in significant flooding throughout the city and surrounding area (OECD, 2013b).

2. Although, to the extent that outdoor recreation on water bodies is excludable, this would be a quasi-private good (Hanemann, 2006).

3. In this report, we define a "water entitlement" as the entitlement to abstract and use water from a specified resource pool as defined in the relevant water plan. In some countries, this may be referred to as "water rights", "water users' rights", "water contracts", an abstraction license or permit.

4. This section is based on a personal communication with Annukka Lipponen and Bo Libert of the UNECE, 30 October 2014.

5. The Helsinki Rules were later superseded in 2004 by the Berlin Rules on Water Resources,

6. This Arrangement and the Convention that replaced it in 2008 remain exceptions.

References

Bjornlund, H. and P. Rossini (2007), "An Analysis of the Returns from an Investment in Water Entitlements in Australia", *Pacific Rim Property Research Journal*, Vol. 13(3), pp. 344-60.

Brisbane Declaration (2007), 10th International River Symposium and International Environmental Flows Conference, Brisbane, Australia, 3-6 September 2007, *www.eflownet.org/download_documents/brisbane-declaration-english.pdf* (accessed 19 March 2014).

Briscoe, J. (1996), "Water as an economic good: The idea and what it means in practice", in *Proceedings of the World Congress of the International Commission on Irrigation and Drainage*, Cairo, Egypt, September.

DWLBC (2006a), "Volumetric Conversion in the South East of South Australia: Calculation of the Delivery Component and Bridging Volume", *DWLBC Report 2006/34*, Department of Water, Land and Biodiversity Conservation, Government of South Australia, *www.waterconnect.sa.gov.au/Content/Publications/DEWNR/dwlbc_report_2006_34.pdf*.

DWLBC (2006b), "Volumetric Conversion in the South East of South Australia: Calculation of Specialised Production Requirements", *DWLBC Report 2006/31*, Department of Water, Land and Biodiversity Conservation, Government of South Australia, *www.waterconnect.sa.gov.au/Content/Publications/DEWNR/dwlbc_report_2006_31.pdf*.

Getzler, J. (2004), *A History of Water Rights at Common Law*, Oxford Studies in Modern Legal History, Oxford University Press.

Griffin, R. et al. (2013), "Introduction: Myths, principles and issues in water trading", in J. Maestu (ed.), *Water Trading and Global Water Scarcity: International Experiences*, RFF Press Water Policy Series, RFF Press, New York, NY, United States.

Hanemann, M. (2006), "The Economic Conception of Water", in Rogers, P.P., M.R. Llamas and L. Martinez-Cortina (eds.), *Water Crisis: Myth or Reality?*, Taylor & Francis plc., London.

Kolstad, C. (2000), *Environmental Economics*, Oxford University Press, New York.

Marshall, G.R. (2005), *Economics for Collaborative Environmental Management: Regenerating the Commons*, Earthscan, London.

Mechlem, K. (2012), "Groundwater Governance: A Global Framework for Country Action", *Thematic Paper 6: Legal and Institutional Frameworks*, Rome, *www.groundwatergovernance.org/fileadmin/user_upload/groundwatergovernance/docs/Thematic_papers/GWG_Thematic_Paper_6.pdf* (accessed 2 November 2014).

OECD (2013a), *Water and Climate Change Adaptation: Policies to Navigate Uncharted Waters*, OECD Studies on Water, OECD Publishing, Paris, *http://dx.doi.org/10.1787/9789264200449-en*.

OECD (2013b), *Water Security for Better Lives*, OECD Studies on Water, OECD Publishing, Paris, *http://dx.doi.org/10.1787/9789264202405-en*.

Olmstead, S. (2010), "The Economics of Managing Scarce Water Resources", *Review of Environmental Economics and Policy*, Vol. 4(2), pp. 179-198, *http://dx.doi.org/10.1093/reep/req004*.

Ostrom, E. (2003), "How Types of Goods and Property Rights Jointly Affect Collective Action", *Journal of Theoretical Politics*, Vol. 15(3), pp. 239-270.

Ostrom, E. (1992), *Crafting Institutions for Self-governing Irrigation Systems*, ICS Press, San Francisco, CA, United States.

Otterman, E. and S. Koeppel (2014), "The UNECE Water Convention and its Program of Adaptation to Climate Change in Transboundary Basins", in Sanchez, J.C. and J. Roberts (2014), *Transboundary Water Governance: Adaptation to Climate Change*, International Union for Conservation of Nature, Gland, pp. 159-174, *https://portals.iucn.org/library/efiles/documents/IUCN-EPLP-no.075.pdf* (accessed 2 November 2014).

Pearce, D. and E. Özdemiroglu (1997), *Integrating the Economy and the Environment: Policy and Practice*, Commonwealth Secretariat, London, United Kingdom.

Poff, N.L. et al. (2009), "The ecological limits of hydrologic alteration (ELOHA): a new framework for developing regional environmental flow standards", *Freshwater Biology*, No. 55, Blackwell Publishing Ltd, pp. 147-170.

Samuelson, P.A. (1954), "The pure theory of public expenditure", *Review of Economics and Statistics*, Vol. 36(4), November, pp. 387-389.

South East Natural Resources Management Board (2011), *Guide to the Development and Contents of the 2009 Padthaway Water Allocation Plan*.

UNECE (2011), "Drainage basins of the North Sea and Eastern Atlantic", *Second Assessment of Transboundary Rivers, Lakes and Groundwaters*, United Nations Economic Commission for Europe, New York and Geneva, *www.unece.org/fileadmin/DAM/env/water/publications/assessment/English/ K_PartIV_Chapter7_En.pdf* (accessed 2 November 2014).

UNECE (2009), *River Basin Commissions and Other Institutions for Transboundary Water Cooperation*, United Nations, New York and Geneva.

Whittington, D. et al. (2013), "The Economic Value of Moving Toward a More Water Secure World", *TEC Background Papers*, No. 18, Global Water Partnership Technical Committee, Global Water Partnership, Stockholm.

Wolf, A. (1999), "Criteria for Equitable Allocations: The Heart of International Water Conflict", *Natural Resources Forum*, Vol. 23(1), February 1999, pp. 3-30, *www.transboundarywaters.orst.edu/publications/ allocations* (accessed 18 December 2014).

Wouters, P. et al. (2005), "Sharing Transboundary Waters – An Integrated Assessment of Equitable Entitlement: The Legal Assessment Model", *Technical Documents in Hydrology*, No. 74, UNESCO, Paris, *http://unesdoc.unesco.org/images/0013/001397/139794e.pdf* (accessed 2 November 2014).

Young, M. (2013), *Improving Water Entitlement and Allocation*, background paper for the OECD project on water resources allocation (unpublished).

Chapter 3

The current water allocation landscape

This chapter presents the main findings from the OECD Survey of Water Resources Allocation covering 37 examples of allocation regimes from 27 OECD and key partner countries (BRIICS as well as Colombia, Costa Rica, and Peru). The survey captured information on the current design and functioning of allocation regimes. Overall, the survey provides a solid basis to identify opportunities to improve the performance of current allocation arrangements and informed the development of the policy guidance on allocation presented in the "health check" of Chapter 5.

The statistical data for Israel are supplied by and under the responsibility of the relevant Israeli authorities. The use of such data by the OECD is without prejudice to the status of the Golan Heights, East Jerusalem and Israeli settlements in the West Bank under the terms of international law.

Key messages

- Findings from the OECD Survey of Water Resources Allocation indicate that most allocation regimes make use of the elements of allocation design that can encourage a robust system yet operate with significant limitations. There are a number of areas that could be **adjusted to improve performance** of allocation in terms of economic efficiency, environmental sustainability and social equity.

- There can be **ambiguity between various legal regimes** within a given jurisdiction (e.g. customary rights versus rights designated in different laws). This legal "pluralism" is a source of conflict among water users and increases the likelihood of "allocation by litigation" or "allocation by adjudication", a costly and time-consuming process.

- **Environmental flows are not secured** in at least one-quarter of allocation regimes surveyed. Only 57% of allocation regimes surveyed report **accounting for the potential impacts of climate change** in their allocation arrangements.

- **A sequence of priority uses** is clearly established in nearly all allocation regimes surveyed. Water for "essential uses" (e.g. drinking water, water for national security purposes) usually figures among the highest priority uses, as would be expected. **Water for the environment is rarely among the highest priorities** in times of scarcity and often figures among the lowest.

- While a significant majority of allocation regimes (92%) have a clear definition on the **limit (or "cap") on consumptive use**, these limits may not be respected in practice and only a few rely on flexible limits.

- **All allocation regimes** surveyed report having **legally defined entitlements** to access to water granted either to individuals, to collective bodies, or both. There is clearly scope to broaden the application of **abstraction charges**.

- Most of the allocation regimes surveyed permit some form of **trading, leasing or transferring water entitlements among users**. However, a wide variety of conditions are placed on such transactions, increasing transaction costs and limiting the extent of trade that occurs in practice.

Despite its importance for water management and for reaping the benefits of water resources, a solid evidence base of how water allocation works across a range of contexts is lacking. This chapter helps to fill this gap by providing an overview of water allocation in practice, based on 37 examples of allocation regimes from 27 OECD and key partner countries (BRIICS as well as Colombia, Costa Rica, and Peru). It is based on information gathered via the OECD Survey of Water Resources Allocation.[1] The survey captured information relevant for understanding the current design and functioning of allocation regimes, summarised in Box 3.1. This chapter presents the main findings from the survey.

Box 3.1. **OECD Survey of Water Resources Allocation**

The survey captures key elements of allocation regimes, including:

- *General contextual information at national level:* To provide the overarching institutional and legal context within which water allocation regimes operate and to signal recent efforts to identify areas where water scarcity is becoming a problem. Respondents were also requested to signal any recent or ongoing reforms of allocation regimes.

- *Key elements of the allocation regime:* To provide a detailed view of the functioning of specific allocation regimes, the questionnaire captured information about specific examples. In countries where there are a number of different approaches to water allocation (for example, different allocation regimes for surface or groundwater, or variations in allocation from one province/state/river basin to another) several examples from each country could be provided. The specific information collected relates to:

  - *Physical characteristics of the water system concerned.* This includes variability of flow, the nature of existing infrastructure (if any), as well as an estimation of the relative share of water uses.

  - *How the available resource pool is defined.* This includes identifying if there is a clear limit on consumptive use and if so, how this is defined. It also includes information about how a number of factors are taken into account in determining the available resource pool, including environmental flows, base flow requirements, climate change, etc.

  - *How users access water.* This section documents if and how water entitlements are defined and administered. It covers the main types of arrangements: informal, administered regimes (priority ranking), based on economic instruments (prices, markets).

  - *How access to water works in practice.* Building on the previous section, this includes more detailed information on the characteristics of entitlements (e.g. possibility to trade, lease or transfer) and the possibility to restrict new entrants.

  - *How exceptional circumstances are managed.* This concerns unplanned events or "shocks" that negatively impact on the water resource. It captures information on how such shocks are defined and managed, in terms of the implications for water allocation.

  - *How access is monitored and enforced.* This covers whether and how withdrawals for various categories of users are monitored and the sanctions for non-compliance (if any).

Countries were asked to provide general contextual information (legal and institutional) at the national level as well as flag recent (within the past 10 years) or ongoing reforms of allocation regimes. They were also asked to provide in-depth information about how specific allocation regimes function. In some countries, there is only one allocation regime that prevails across the entire territory. In many other countries, there exist a number of different allocation regimes. For instance, allocation may differ from one province/state/river basin to another. Allocation may also differ for surface water and ground water systems. Therefore, countries were asked to provide at least one or more examples of different allocation regimes. This analysis is based on the specific examples in Table 3.1.[2]

Table 3.1. **Examples of water allocation regimes**

	Allocation regime example(s) provided
Australia	Murray-Darling Basin.
Austria	Surface and ground water systems in Austria (referred to under the Water Act).
Brazil	São Francisco Basin. São Marcos Basin.
Canada	Province of Alberta. Province of Manitoba. Province of Newfoundland and Labrador. Province of Nova Scotia. Province of Prince Edward Island. Province of Quebec. The Yukon Territory.
Chile	Limarí River Basin. Maipo River's, 1st Section; Santiago de Chile.
The People's Republic of China (hereafter "China")	Yellow River Basin.
Colombia	Ubaté-Suárez River Basin.
Costa Rica	National scale.
Czech Republic	Water use system in the Czech Republic (referred to under the Water Act).
Denmark	National scale (4 river basin districts, 23 sub-basin districts).
Estonia	National scale.
France	Organisations for the collective management of irrigation (national scale).
Hungary	Tisza-Körös Valley Management System.
Israel	Three types of regimes at national scale: ● Wastewater treatment. ● Desalination. ● Municipal/regional water corporations.
Japan	Tone-Gawa River System, five prefectures: Ibaraki, Tochigi, Gunma, Chiba, Tokyo.
Korea	Water systems in Korea (referred to in the River Act).
Luxembourg	National scale.
Mexico	National scale (13 hydrological regions).
Netherlands	The Dutch Polder System (in the western part of the Netherlands).
New Zealand	Waikato Region.
Peru	Parón River's Sub-Basin.
Portugal	Tejo River Basin.
Slovenia	National scale.
South Africa	Inkomati, Jan Dissels and Mhlatusa River Basins.
Spain	National scale.
Switzerland	National scale.
United Kingdom	England and Wales.

Source: See country profiles associated with this publication at *www.oecd.org/environment/water-resources-allocation-9789264229631-en.htm.*

Examining the survey results in context

The information captured in the OECD survey provides a varied view of the current allocation landscape across a range of countries with diverse water endowments, different types of challenges relating to freshwater supply and demand, and varying legal, institutional and policy settings. Although this set of responses provides a diversity of allocation examples, to put the results into context it is important to note several factors that influenced the type of examples collected and the extent to which they are likely to be representative of broader conditions for allocation worldwide.

First, the questionnaire requested specific and detailed information relating primarily to *formal* arrangements for allocation. Examples of informal allocation were therefore unlikely to be captured, although they may be common in many countries. Countries that have traditionally faced water scarcity issues are, in general, more likely to have put in place formal policy responses to address this. The comprehensive water policy reforms in the Murray-Darling Basin are a good example. However, traditionally water-abundant countries facing *emerging* scarcity issues, such as the Netherlands, tend to have fewer formal arrangements in place for managing allocation, and hence, less to report via the questionnaire. Finally, the questionnaire captured the legal, institutional and policy arrangements as reflected in formal arrangements, but this survey is not well-suited to capture how these arrangements are implemented in practice. The difference between formal allocation arrangements and how allocation functions in practice could be significant, with important implications for the sustainable use of the resource.

For example, for Korea, the questionnaire response covers allocation arrangements as reflected in the River Act, although there is ambiguity about how the River Act relates to water use claims under Civil Law, which recognises customary rights, but does not define precisely how this should be done. This legal "pluralism" and ambiguity is a source of significant conflict over water allocation in the country. In a similar vein, the response for Japan covers formal allocation arrangements of the Tone-Gawa River System, but it does not cover issues related to water entitlements under customary law. The response for England and Wales reflects current arrangements in the abstraction licensing regime, but does not reflect numerous legacy issues, such as non-time-limited permits, which represent a significant claim on water resources. The main point here is that while this analysis of the current allocation landscape can provide a number of very useful insights, the survey approach and the information captured need to be understood as reflecting certain aspects of allocation, while leaving others to be further explored in other endeavours.

The following sections will examine the survey results for each element of the allocation framework described in Chapter 2, drawing out key findings.

Reforming water allocation regimes

The survey confirmed the assumption that water resources allocation is a dynamic policy area, with a number of countries either currently reforming allocation arrangements or have recently done so (Table 3.2). Seventy-five per cent of countries (including responses from Canadian Provinces and Territories) indicated that their allocation regimes had been recently reformed. Ongoing reforms were flagged by over a majority (62%) of countries (including responses from Canadian Provinces and Territories).

Table 3.2. **Countries with recent or ongoing water allocation reforms**

	Australia	Austria	Brazil	Canada	Chile	China	Colombia	Costa Rica	Denmark	France	Israel	Luxembourg	Mexico	Netherlands	New Zealand	Peru	Portugal	Slovenia	Spain	South Africa	Switzerland	United Kingdom
Recent reforms	•	•	•	•	•	•	•	•	•	•	•	•	•		•	•	•	•	•	•		•
Ongoing reforms	•			•	•	•	•	•		•	•		•	•	•	•		•	•	•	•	•

Note: For Canada: recent reforms were flagged by Manitoba while ongoing reforms were flagged by Alberta, Quebec and the Yukon Territory.
Source: See country profiles associated with this publication at *www.oecd.org/environment/water-resources-allocation-9789264229631-en.htm*.

Environmental improvement or protection was the most frequently cited driver of both recent and ongoing reforms of allocation regimes. For ongoing reforms, the second most frequently cited driver was concerns about equity in access to water. For recent reforms (in the past 10 years), economic development was the second most frequently cited driver, following environment improvement or protection. Drivers of recent and ongoing reforms cited by countries are summarised in Figure 3.1 (multiple responses were possible).

Figure 3.1. **Drivers of recent and ongoing reforms of water allocation regimes**

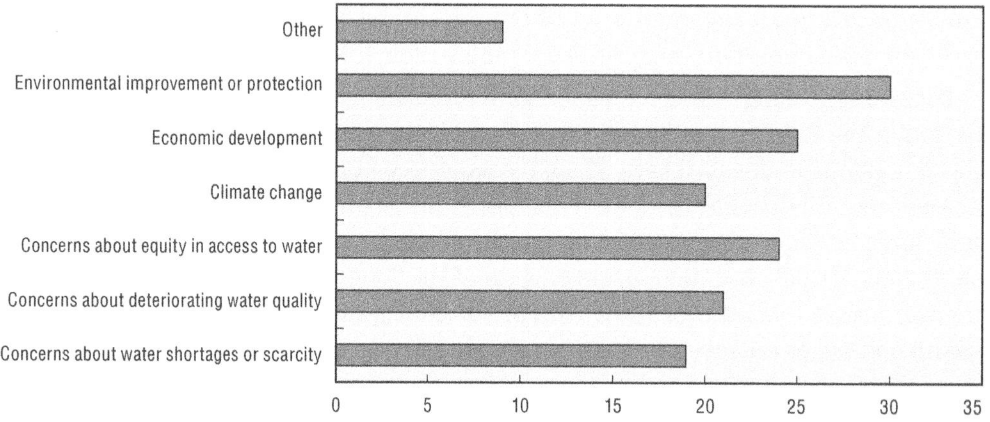

Source: See country profiles associated with this publication at *www.oecd.org/environment/water-resources-allocation-9789264229631-en.htm*.

In addition to the drivers of reform indicated in Figure 3.1, countries provided additional reasons for pursuing reforms of allocation regimes. For example, the main driver cited by Mexico was to achieve better control of water resources and to discourage illegal practices. New Zealand noted that addressing Māori rights and interests regarding fresh water was among the drivers of both recent and ongoing reforms.

General contextual information for allocation

General contextual information provides useful background for understand the setting within which allocation regimes operate. Information about the legal setting allows for an understanding of the ownership of water resources (ground and surface water). Insight into the institutions primarily responsible for water allocation and their main responsibilities allow for a basic understanding of the organisations responsible for allocation. Countries also provided information about whether they were tracking water scarcity.

Legal setting for water resources

When asked how the ownership of water resources (both surface and groundwater) was legally defined (if at all), the large majority of countries (85% for surface water, 77% for ground water) indicated that they are publicly owned (Figures 3.2 and 3.3). It is important to note that "ownership" here refers to ownership of the resource itself, *not* the entitlement or right to use the resource (discussed in a later section). Instances of privately owned water resources usually relate to ground water. Estonia is the sole exception where surface water may be either publicly or privately owned.

Figure 3.2. **Ownership of groundwater resources**

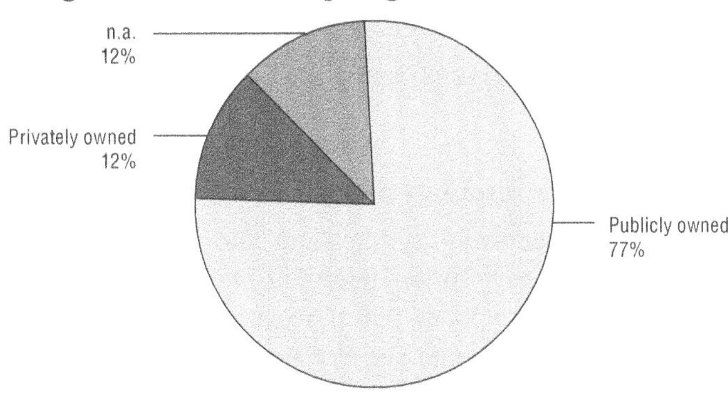

Note: Does not include Switzerland.
Source: See country profiles associated with this publication at *www.oecd.org/environment/water-resources-allocation-9789264229631-en.htm.*

Figure 3.3. **Ownership of surface water resources**

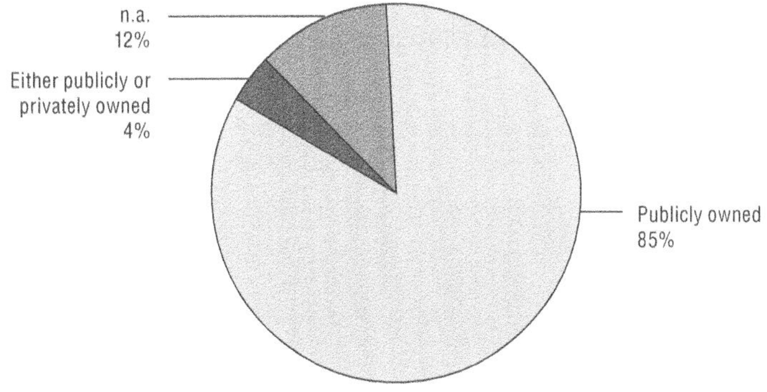

Note: Does not include Switzerland.
Source: See country profiles associated with this publication at *www.oecd.org/environment/water-resources-allocation-9789264229631-en.htm.*

In the case of groundwater, the resource may be privately owned by the owner of the property overlying it. Austria is a case where groundwater is privately owned. Japan and Portugal indicated that groundwater may be privately owned in some cases. In some countries, it is explicitly stated that water resources are not the subject of legal ownership, either public or private (indicated as "n.a." in Figures 3.2 and 3.3). In these cases, water resources in these cases may be designated as *res nullius*, or "ownerless property" in legal terms.

To illustrate the various legal doctrines applied to water resources, Table 3.3 summaries diverse examples.

Table 3.3. **Examples of legal basis/doctrine applied to water resources**

Country/region	Legal basis/doctrine applied to water resources
Australia	Owned by the Crown.
France	Ground and surface water are defined as the "Common Heritage of the Nation".
Japan	Surface water cannot be made the subject of private ownership.
Luxembourg	Surface and groundwater is legally defined as *Res nullius* ("ownerless" property).
Mexico	Property of the Nation.
Province of Manitoba, **Canada**	Owned by the Provincial Crown.
United Kingdom	Water is not "owned" as such, but the land adjacent to it or overlying it can be.

Source: See country profiles associated with this publication at *www.oecd.org/environment/water-resources-allocation-9789264229631-en.htm.*

Institutional setting for water resources allocation

In most countries, the responsibility for water allocation is shared among several institutions, at various levels of government (national, state/provincial/regional, basin, and local). Key responsibilities relate to policy, planning, issuing water entitlements, as well as monitoring and enforcement. As summarised in Figure 3.4, slightly fewer than half (48%) of countries indicated a role for the Ministry of Environment in water allocation. Even fewer (30%) indicated a role for a basin authority (Figure 3.5). Countries often indicated that a sub-national authority (or authorities), such as a department or agency responsible for water, figured among the institutions with responsibility for water resources allocation.

Figure 3.4. **Percentage of responses indicating a role
for the Ministry of Environment**

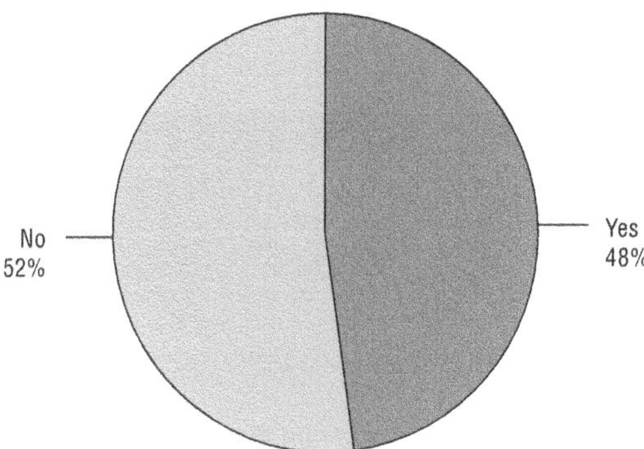

Note: Considered responses indicating Ministry of Environment at the national level (does not consider sub-national environmental authorities/departments).
Source: See country profiles associated with this publication at *www.oecd.org/environment/water-resources-allocation-9789264229631-en.htm.*

Figure 3.5. **Percentage of responses indicating a role for a basin authority**

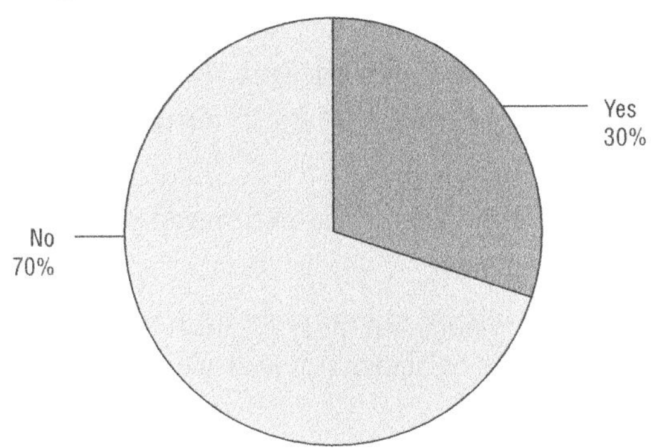

Source: See country profiles associated with this publication at *www.oecd.org/environment/water-resources-allocation-9789264229631-en.htm.*

Tracking water scarcity

In a majority of countries (74%), a mapping exercise has been undertaken to identify areas where surface and ground water scarcity is becoming a problem (Table 3.4). Links to specific studies can be found in the country profiles.

Table 3.4. **Countries recently assessing water scarcity**

	Australia	Austria	Brazil	Canada	Chile	China	Colombia	Costa Rica	Czech Republic	Denmark	Estonia	France	Hungary	Israel	Japan	Korea	Luxembourg	Mexico	Netherlands	New Zealand	Peru	Portugal	Slovenia	South Africa	Spain	Switzerland	United Kingdom
Tracking scarcity	•	•	•	•	•	•	•	•			•	•	•			•		•	•	•	•		•	•	•		•

Note: For Canada: the Provinces of Alberta, Quebec, Nova Scotia, and Newfoundland and Labrador indicated that an assessment had been undertaken, while Manitoba and the Yukon Territory indicated that no assessment had been undertaken. For Israel: an assessment is now under preparation. For New Zealand: the assessment recently undertaken only covers surface water.
Source: See country profiles associated with this publication at *www.oecd.org/environment/water-resources-allocation-9789264229631-en.htm.*

The information summarised above on allocation reforms as well as the general contextual information (legal and institutional setting, as well as information about tracking water scarcity) was captured at the national level.[3] Respondents to the survey also provided specific and detailed information about one or more allocation regime examples (as detailed in Table 3.1). The remaining sections of this chapter summarise information from these 37 examples.[4]

Understanding the physical features of the water resource and demand profile

An understanding of the physical characteristics of the water system, the demand profile of users in that system and the current status of the resource are pre-requisites for designing an appropriate allocation regime. Physical characteristics include the variability of flow, connectively with ground and surface water bodies, the nature of existing infrastructure (if any), the extent to which the flow rate can be controlled or not, as well as an estimation of the relative share of water uses. The current status of the water system is indicative of the need for various degrees of elaboration of an allocation regime.

Current status of water system

Generally, the greater the pressure on the resource, the greater the need for formal control of the resource and well-defined and flexible allocation arrangements. Respondents were asked to characterise the status of the water system in one of three categories (Figure 3.6):

- **Over-used:** existing abstractions exceed the estimated proportion of the resource that can be taken on a sustainable basis.

- **Over-allocated:** current use is within sustainable limits but there would be a problem if all legally approved entitlements to abstract water were actually used.

- **Neither** over-allocated nor over-used.

Figure 3.6. **Proportion of water allocation examples by current status of water systems**

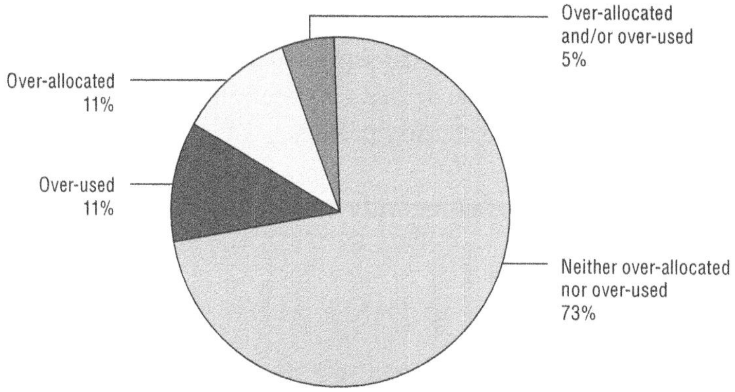

Over-allocated and/or over-used
5%

Over-allocated
11%

Over-used
11%

Neither over-allocated nor over-used
73%

Source: See country profiles associated with this publication at *www.oecd.org/environment/water-resources-allocation-9789264229631-en.htm.*

The majority (73%) of allocation examples are currently considered neither over-allocated nor over-used. Eleven per cent are considered over-allocated, 11% over-used. Two examples, the Maipo River, 1st Section, in Chile, and various regions in Spain are considered either over-allocated and/or over-used. For the allocation regimes considered over-allocated or over-used, respondents reported a number of different measures being undertaken to address this issue. For instance, in New Zealand's Waikato region, measures are in place to phase out the exceedance of allocable flows by ceasing new allocations, encouraging voluntary reduction, and promoting augmentation of water supplies. In Australia, a range of programmes have been initiated to recover water entitlements in regions where water abstractions are in excess of the sustainable diversion limit. Once recovered, these entitlements are transferred to the Commonwealth Environmental Water Holder and managed for environmental purposes.[5]

Degree of regulation of flow rate of water resources

To provide an understanding of the extent to which the flow rate of a given water system can be managed or controlled, respondents were asked to classify the water system by the degree it is regulated. Water systems could be characterised as:

- **Fully regulated:** the flow rate can be controlled fully.

- **Partially regulated:** the flow rate can be controlled to some extent.

- **Not regulated:** the flow rate cannot be controlled.

For some examples, particularly those that cover an extensive water system or reflect responses at the national scale, respondents indicated that the degree of regulation of the system depends on the particular location. Responses are summarised in Figure 3.7.

Figure 3.7. **Degree of regulation of water system**

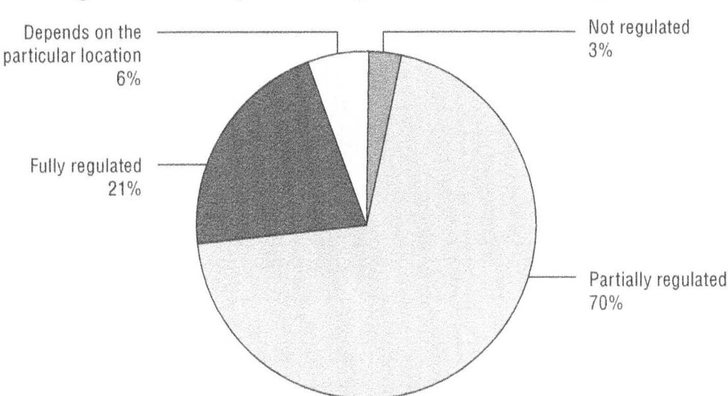

Note: Ubaté-Suárez River Basin (Colombia), Mexico, and South Africa are not included.
Source: See country profiles associated with this publication at *www.oecd.org/environment/water-resources-allocation-9789264229631-en.htm.*

The flow rate of water system is partially controlled in 70% of allocation examples, indicating the presence of the technical capacity and infrastructure to facilitate allocation. In the Murray-Darling Basin, Australia and in England and Wales, the flow rate can be either fully, partially or not at all, depending on the particular location. In only one of the examples collected, the Yukon Territory in Canada, the water system is completed unregulated.

Consumptive use profile

The breakdown across categories of users in a water system provides useful information on the particular demand profile served by an allocation regime. Both consumptive and non-consumptive uses are relevant. To capture the breakdown across consumptive uses, respondents were asked to provide a general estimate of the percentage of mean annual inflow/recharge[6] that is consumed across specific categories (agriculture, domestic, industrial, energy production (not including hydropower), environment (evapotranspiration), transfer to the sea or another system, or "other". Figure 3.8 summaries the proportion of examples according to the dominant type of water user.

Agriculture is the dominant water user in half of the 26 examples analysed. The Czech Republic, Estonia, France and Mexico indicated energy production as the dominant water user. In the São Francisco and São Marcos Basins in Brazil, and the Province of Manitoba in Canada most of the water is available for environmental purposes (with a portion consumed for evapotranspiration). However, this does not preclude tensions regarding water use between various uses in these basins due to the specific timing and location of water use.

Figure 3.8. **Proportion of water allocation examples according to dominant type of water use, per category**[1]

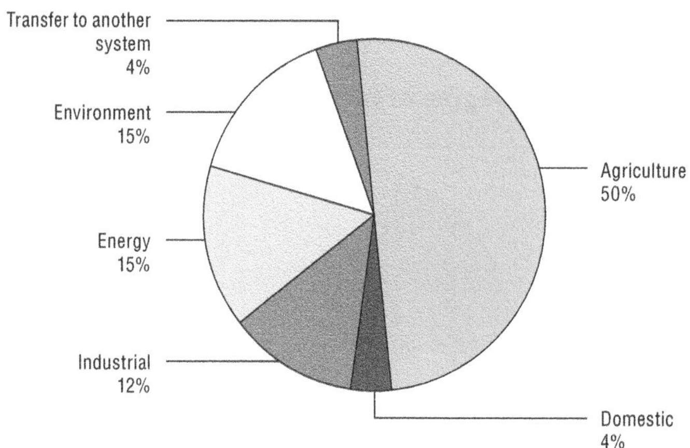

Note: A number of examples were excluded: Murray-Darling Basin (Australia), Newfoundland and Labrador, Nova Scotia, Quebec (Canada), Maipo River's, 1st Section (Chile), Denmark, the Dutch Polder System in the western part (the Netherlands) Tejo River Basin (Portugal), Parón River's Sub-basin (Peru), Slovenia, and England and Wales (United Kingdom).

1. Note that in the questionnaire, "water use" by the environment was considered evapotranspiration.

Source: See country profiles associated with this publication at *www.oecd.org/environment/water-resources-allocation-9789264229631-en.htm.*

Significant non-consumptive uses

Significant non-consumptive uses were indicated in 78% of the examples. Of the examples reporting significant non-consumptive uses, the proportion of examples reporting various types of non-consumptive use are summarised in Figure 3.9. Hydropower is clearly a dominant type of non-consumptive use, indicated in 86% of the allocation examples that report significant non-consumptive use. Transportation and navigation was the second most frequently cited significant non-consumptive use (multiple responses were possible).

Figure 3.9. **Proportion of water allocation examples indicating significant non-consumptive use, by type**

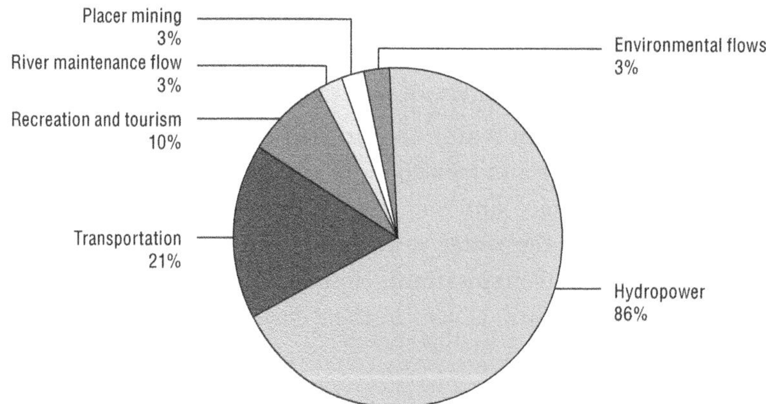

Source: See country profiles associated with this publication at *www.oecd.org/environment/water-resources-allocation-9789264229631-en.htm.*

Defining the available ("allocable") water resource pool

As described in Chapter 2, the definition of the available resource pool that can be allocated for consumptive use is a key element of a robust allocation regime. This definition determines the repartition between water available for *in situ* uses and diverted uses. It can be used to set a clear limit, or "cap", on abstractions, which can be adjusted over time as circumstances change. This section examines important "system level" information, summarising how allocation regimes define the limits on abstraction, whether environmental flows (e-flows) are secured and whether various other factors are taken into account in the definition of the available resource pool to contribute to the hydrological integrity of the regime.

Defining limits on water abstraction

Respondents were asked if there is a clear definition of the limit on consumptive use, and if so, how this limit is defined, among the following options (responses summarised in Figure 3.10):

- A limit on the **volume of water** that can be abstracted.

- A limit on the **proportion** (e.g. percentage of flow) of water that can be abstracted.

- Restrictions on **who** can abstract water (but no limit on how much water can be abstracted).

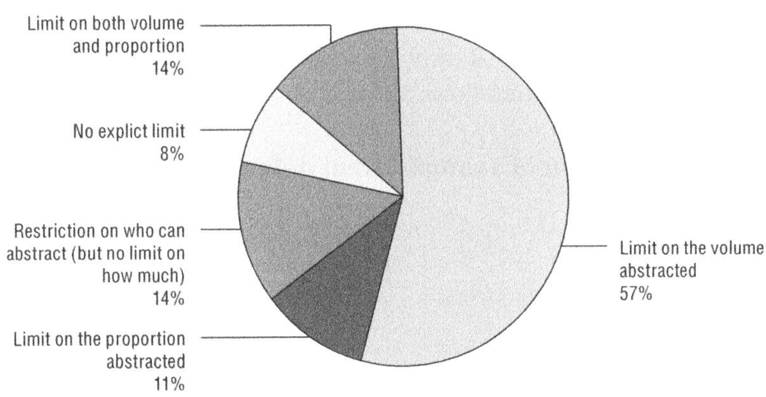

Figure 3.10. **Proportion of examples according to type of limit on water abstraction (if any)**

Source: See country profiles associated with this publication at *www.oecd.org/environment/water-resources-allocation-9789264229631-en.htm.*

A significant majority of allocation regimes (92%) have an explicit limit on abstraction for consumptive use. Only three examples report having no explicit limit on abstraction: the Czech Republic, the Netherlands and the Yukon Territory in Canada. A limit based solely on the volume abstracted is the most common type of definition with 57% of examples using this approach. Only a few examples (11%) use a limit based solely on the proportion abstracted. A hybrid approach is taken by 14% of examples, with limits that are set both in terms of volume and proportion. Fourteen per cent of examples have a restriction on who can abstract water (but no limit on how much).

Of those allocation regimes with an explicit limit on abstractions, 42% indicated that the amount of water available for consumptive use in the resource pool is linked to a river basin management plan and 33% indicated that it is linked to another planning document. Around a fifth of responses (21%) indicated that the limit is not linked to any planning document. For those examples that have linked the limit on consumptive use to an official planning document (river basin management plan or otherwise), just over half (56%) indicated that the document was a statutory instrument that must be followed, while the remainder indicated that the plan was considered a guiding document.

Defining environmental flows

A significant majority (76%) of examples indicated that environmental flows are defined (Figure 3.11). A wide range of methodologies to do so was reported. For example, in Israel, in some places a minimum quota of water has been set aside and must be allocated to ecosystems. In Slovenia, the ecologically acceptable flow is set depending on the type of water use and type of ecological needs. In England and Wales, environmental flow indicators are used to determine the flows required by the environment by particular ecosystems. In Portugal, minimum environmental flows are determined on a case by case basis. In China, the warning-level river flow against the drying out of a downstream river course shall not fall below 200 cm^3/second at Xiaheyan hydrological stations. In the Murray-Darling Basin, Australia, the Basin Plan limits water use at environmentally sustainable levels by determining long-term sustainable diversion limits for both surface and groundwater resource. A key component of the Basin Plan is the environmental watering plan, which co-ordinates all environmental watering across the basin. The Plan also contains a water quality and salinity management plan and water quality targets which influence how environmental flows and the water resources are managed.

Figure 3.11. **Proportion of examples that defined environmental flows**

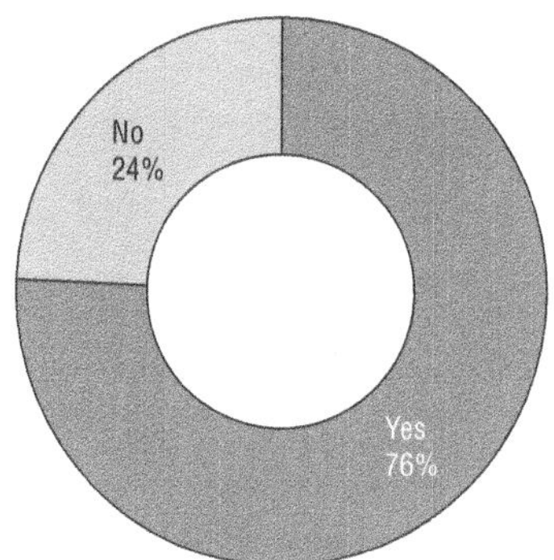

Source: See country profiles associated with this publication at *www.oecd.org/environment/water-resources-allocation-9789264229631-en.htm.*

Of the examples indicating that environmental flows are taken into account, 82% take freshwater biodiversity into account in the definition of e-flows and 64% take terrestrial biodiversity into account (Figures 3.12 and 3.13). For example, in France, the minimum biological flow and the reserve flow required are based on the observation of ecological needs.

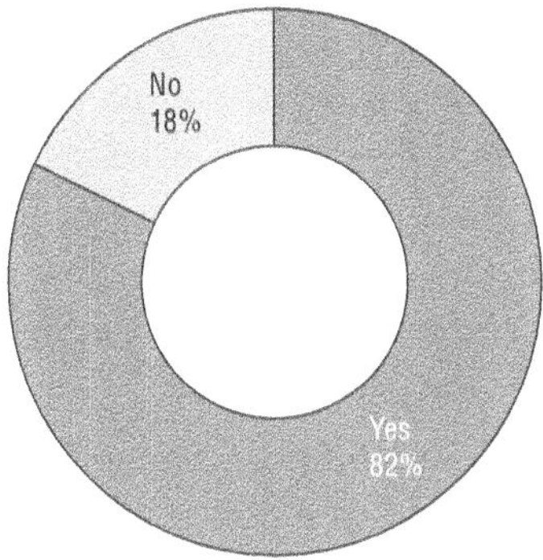

Figure 3.12. **Proportion taking freshwater biodiversity into account in the definition of e-flows**

Source: See country profiles associated with this publication at *www.oecd.org/environment/water-resources-allocation-9789264229631-en.htm.*

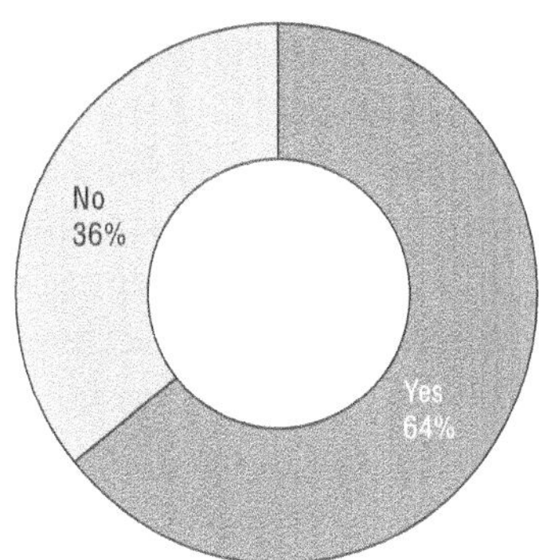

Figure 3.13. **Proportion taking terrestrial biodiversity into account**

Source: See country profiles associated with this publication at *www.oecd.org/environment/water-resources-allocation-9789264229631-en.htm.*

Factors accounted for in the definition of the available water resource pool

When defining the available resource pool, various factors can be taken into account, to ensure the hydrological integrity of the system and account for variability. Figure 3.14 depicts the proportion of allocation examples that report that key factors are taken into account in the definition of the available resource pool. Specific information about how this is done, such as the methodological approach used, are reflected in country profiles.

Figure 3.14. **Proportion of water allocation examples taking into account various factors in the definition of the available resource pool**

Percentage of responses

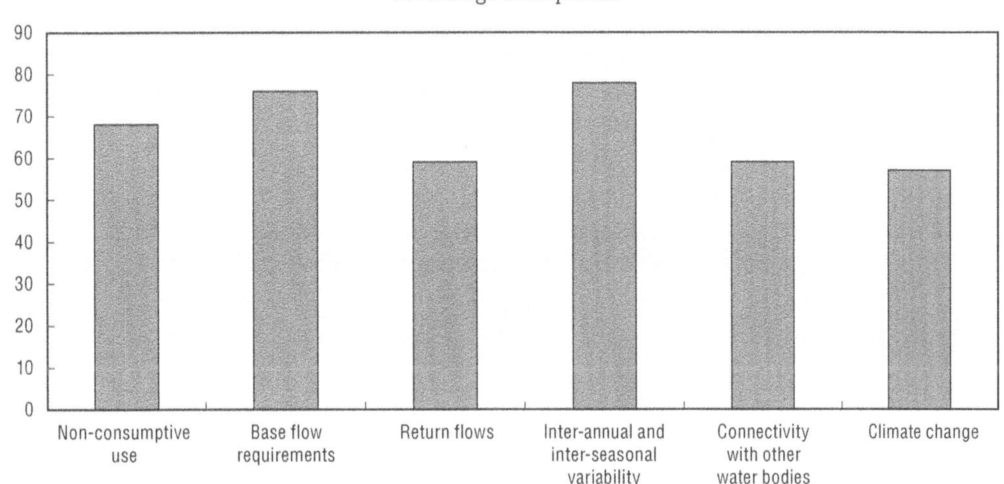

Source: See country profiles associated with this publication at *www.oecd.org/environment/water-resources-allocation-9789264229631-en.htm.*

Inter-annual and inter-seasonal variability is the factor most frequently taken into account (in 78% of the allocation regimes surveyed). For example, in England and Wales, environmental standards take into account inter seasonal activity, as do seasonal abstraction licenses. Inter annual variability can be taken into account via time-limited licences. In Alberta, Canada, before an abstraction license is issued, a review will be undertaken of the seasonal maximum and minimum flows and the license will specify amounts that may be withdrawn based on annual variability.

Base flow requirements are also frequently taken into account in the definition of the available resource pool (in 76% of regimes surveyed). In France, base flow is mandatory and is not included in the volumes defined as abstractable. Recent reforms of freshwater management in New Zealand require all regions to set environmental flows (including an allocation limit and minimum flow) by 2025. Current approaches to base flow vary by region. For example, the Waikato region sets minimum base flows as a percentage of the one in five year 7-day low flow following detailed habitat and river studies.

Non-consumptive uses are considered as a factor in defining the available resource pool in 68% of the examples. In the case of England and Wales, non-consumptive uses are included in the abstraction management system. They require a license and their right to abstract is protected.

Only 57% of allocation regimes report taking into account climate change, about the same proportion that consider return flows and connectivity with other water bodies. Taking climate change into account in the definition of available resources typically involves periodic updating of the scientific basis. For instance, in Austria, the impact of climate change on the availability of resources is investigated scientifically on a regular basis.

How users access water and how this works in practice

To better understand "user level" issues in an allocation regime, detailed information was collected about how water entitlements (referred to in some countries as abstraction licenses or permits, or water use rights) are defined and work in practice.

Legal definition of water entitlements

In all allocation regimes examples collected, with the exception of the Netherlands' polder system in the western part of the country, water users' entitlements to abstract or divert water from the resource pool are legally defined. The majority of examples (89%) allow for legally defined private entitlements (Figure 3.15).

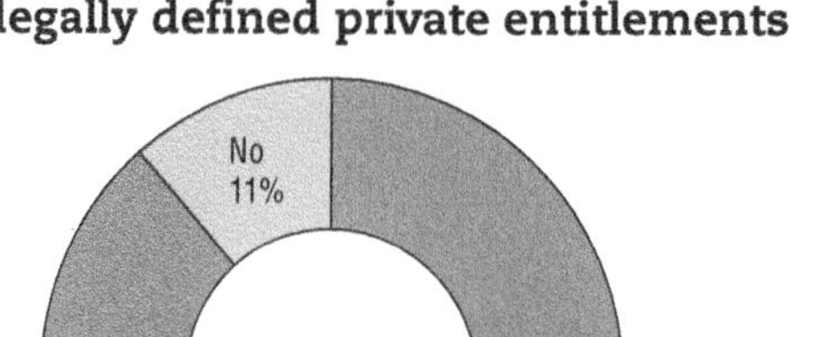

Figure 3.15. **Proportion of water allocation regimes with legally defined private entitlements**

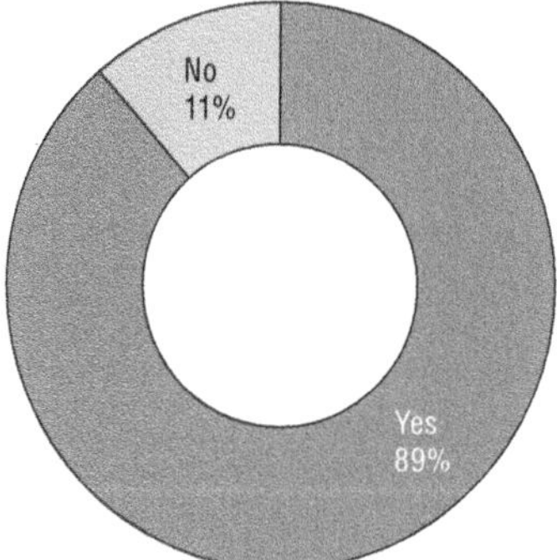

Note: Does not include: the Dutch Polder system in the western part of the country (the Netherlands) and the Inkomati, Jan Dissels and Mhlatuse River Basins (South Africa).
Source: See country profiles associated with this publication at *www.oecd.org/environment/water-resources-allocation-9789264229631-en.htm.*

Private entitlements can take several forms, including an individual entitlement (to an individual person), a collective entitlement (to a group of persons/organisation/city), or an alternative arrangement. Figure 3.16 indicated the number of allocation regimes that report that entitlements are granted to individuals (10 examples) and the number of regimes in which an entitlement can be granted to either an individual or a collective body (21 examples).

Figure 3.16. **Number of water allocation examples by type of entitlement (individual, collective)**

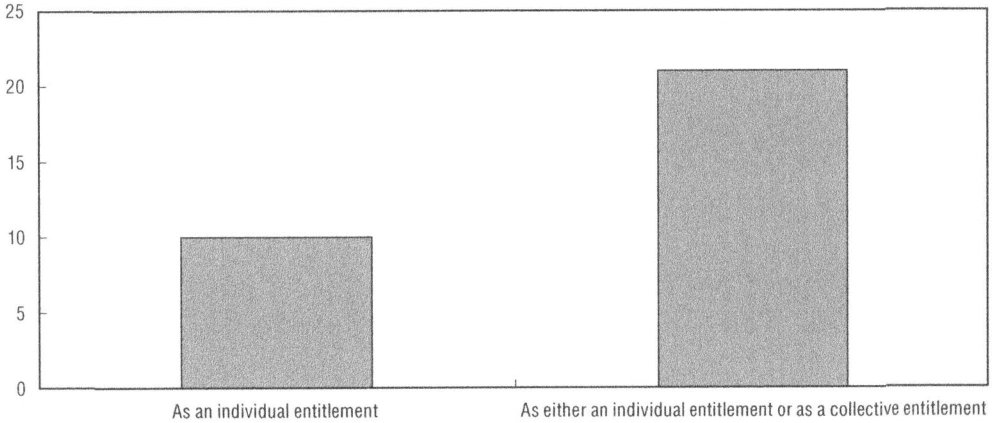

Source: See country profiles associated with this publication at *www.oecd.org/environment/water-resources-allocation-9789264229631-en.htm.*

For allocation regimes where collective entitlements are possible, a number of arrangements to allocate water among specific users were reported. For example, in the Murray-Darling Basin, Australia, for collective irrigator groups with a collective entitlement, the individual entitlement is defined as a share of the collective entitlement based on the rules of association of their membership. For urban authorities providing town water supply, individuals enjoy unlimited supply on a pay for use basis (typically on a full cost basis). Different levels of restriction may be imposed to further limit demand and subsequent use in periods of low allocation to the urban entitlement.

In Alberta, Canada, in the case of collective entitlements, allocation of water among individual users within a group of users is based on a bargaining process and informal trading. In the Yellow River Basin, China, collective entitlements are assigned to an institution representing water users. Irrigation districts and public water companies access water to consume by paying a fee. In some irrigation districts, authorities assign water abstraction rights to clients under a permit system. For Costa Rica, in the case of collective entitlements, the Ministry of Energy and Environment grants a concession to each Society of Water Users according to the Water Law. These societies have the authority to decide internally the form of water distribution amongst their members through agreements of the general assembly of members, or through their own regulations. In Spain, there are both individual and collective entitlements. Collective entitlements may be granted to Water Users Associations or Irrigators Communities, for instance. Finally, in the case of France, the recently created Single Collective Management Bodies (OUGC) provide a structure and incentives for irrigators to devise their own rules to allocate a set volume of water among themselves at the catchment level. These rules are subject to approval by the Ministry of Ecology, Sustainable Development and Energy.

Nature of water entitlements

Water entitlements can be defined in a number of different ways: as entitlements to abstract water unbundled from property titles, as riparian entitlements linking the access of water to the ownership of adjacent land, or via a system of prior appropriation, where reliability of water access is a function of the date the entitlement was first issued (further discussion in Chapter 2). In the allocation regimes surveyed, a majority reported that water entitlements were unbundled from property titles (21 of 29 examples), with only a few using riparian entitlements or prior appropriation (Figure 3.17). Four examples indicated that entitlements were defined as a combination of these various approaches or depended the particular context.

Figure 3.17. **Nature of water users' entitlements**

Number of examples

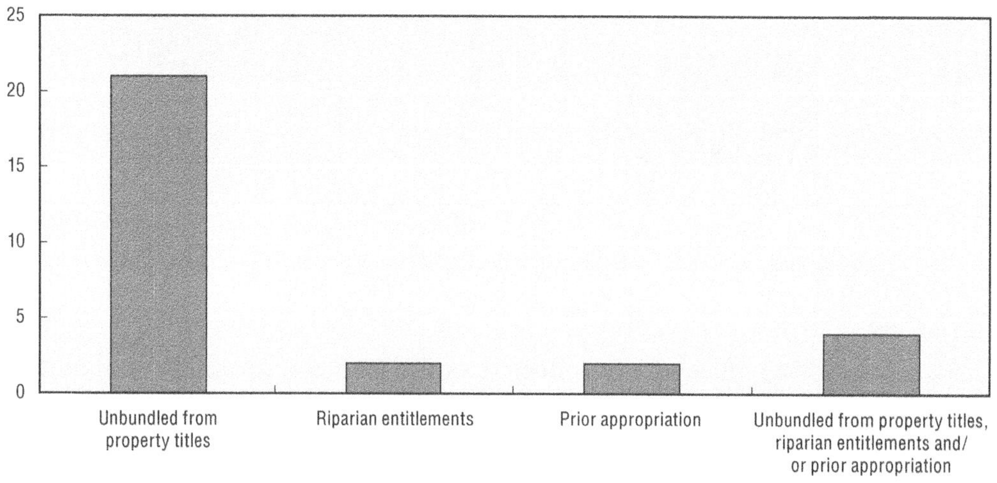

Note: Does not include: Nova Scotia, Prince Edward Island (Canada), Ubaté-Suárez River Basin (Colombia), wastewater reuse, large scale desalination and municipal/regional water corporations (Israel), the Dutch polder system in the western part of the country (the Netherlands), and Switzerland.
Source: See country profiles associated with this publication at *www.oecd.org/environment/water-resources-allocation-9789264229631-en.htm.*

Period of time water entitlement granted for

In most cases, water entitlements are time bound, either with or without an expectation of renewal. The majority of the allocation regimes surveyed grant water users' entitlements for a given number of years, with the expectation of periodic renewal. However, seven allocation regimes in four countries granted entitlements in perpetuity. In Chile (both the Limarí River Basin and the Maipo River, 1st Section) entitlements are granted in perpetuity without conditions relating to beneficial use. In the Murray-Darling Basin, Australia, as well as the three examples from Israel (wastewater re-use, large scale desalination and local/regional water corporations), entitlements are granted in perpetuity, but conditional upon beneficial use. In Peru, entitlements are also granted in perpetuity, but conditional on continuity of use.

Figure 3.18 summarises the incidence of examples indicating a fixed period of time for entitlements (a given number of years) versus the incidence of allocation examples granting entitlements in perpetuity. Most of the examples in which entitlements are granted for a certain number of years indicated that there was an expectation of renewal.

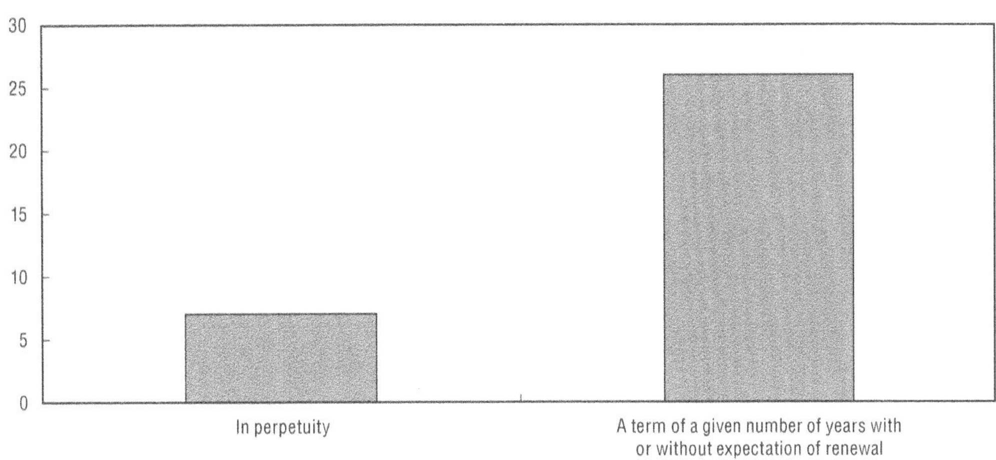

Figure 3.18. **Period of time water entitlement granted for**
Number of examples

Note: Does not include: Prince Edward Island (Canada), the Dutch polder system in the western part of the country (the Netherlands), South Africa (Inkomati, Jan Dissels and Mhlatuse River Basins) and Switzerland.
Source: See country profiles associated with this publication at *www.oecd.org/environment/water-resources-allocation-9789264229631-en.htm.*

Four examples indicated that entitlements were granted for a fixed period of years without the expectation of renewal – Denmark, Estonia, the Tisza-Koros Valley management system in Hungary and the Waikato Region in New Zealand.[7]

The allocation regimes that have time-bound entitlements report a wide range of time periods for which entitlements are granted. The time period is typically dependent on the type of water use or user. Hydropower is afforded the longest duration, by far. Table 3.5 provides a range of illustrations.

Return flows

As discussed previously, "return flows" consist of the water physically withdrawn from a system and returned back to the same or a different body following use. Specifying return flow obligations in water entitlements provides a more accurate reflection of the proportion of water actually consumed and the proportion potentially available for another use (depending on the extent to which the location and quality has changed).

Allocation examples in which return flow obligations were not specified (52%) outnumber those in which return flow obligations were specified (29%), as summarised in Figure 3.19. The remaining 19% of cases specified return flows only in certain cases, or on a case by case basis.

Consequences of non-use of water entitlements

Allocation regimes can place conditions of the exercise of water entitlements and also provisions that allow for entitlements to expire or be revoked in the case that they are not actually used. This is usually to avoid hoarding of entitlements.

In instances where water users' entitlements are not used for a certain period of time, this is usually addressed in one of two ways: either the entitlement is kept in place for the period it was issued (despite not being used) or the entitlement will be lost (applying a "use it or lose it" approach). As summarised in Figure 3.20, 16 allocation regimes reported using a "use it or lose it" system, with 13 reporting that entitlements remain in place for the period they are issued

Table 3.5. **Examples of period of time water entitlements are granted for**

Example	Reference to number of years entitlements are issued for
Australia (Murray-Darling Basin)	In **perpetuity**, conditional on beneficial use.
Austria (surface and ground water systems)	No more than of **90 years** (e.g. for hydropower plants). **12 years:** Maximum term of abstraction for irrigation purposes.
Brazil (São Francisco Basin and São Marcos Basin)	**10 years:** Irrigation of areas up to 2.000 ha; industry with maximum withdraw flow of 1 m³/s; aquaculture; animal consumption; mining; others. **20 years:** Irrigation of areas over 2.000 ha; industry with maximum withdraw flow over 1 m³/s. **35 years:** Dams of flood control or hydropower generation and others hydraulic works; public water supply and sanitation.
Canada (Newfoundland and Labrador)	**5 to 50 years:** depending on user.
Chile (Limarí River Basin and Maipo River's, 1st Section)	In **perpetuity**.
China (Yellow River Basin)	A term of a given number of years (e.g. **5-10**) with the expectation of renewal.
Colombia (Ubaté-Suárez River Basin)	**10 years:** For concessions can be granted for a term not to exceed amount of years. **Up to 50 years:** For public services or the construction of public or social interest.
France (single collective management bodies for irrigation (OUGC))	**Few years to several decades:** Permanent use like drinking water abstraction. **6 months:** Temporary uses (seasonal uses and/or irrigation).
Japan (Tone-Gawa River System)	A term of **10 years** with the exception of 20 years for hydropower generation.
Korea (surface water systems under the River Act)	A term of **10 years** with expectation of periodic renewal.
Luxembourg	A period of **5 to 20 years**, which can be renewed.
Mexico	A term of **5 to 30 years**, with the expectation of periodic renewal.
New Zealand (Waikato Region)	A term of **15 years** without expectation of renewal. However, under the Resource Management Act they can be issued for up to **35 years**. Existing consent holders have the right to have an application for a new permit for the same activity to be considered before other applicants.
Peru	In **perpetuity**, but conditional upon continuity of activity.
Spain	A term of no more than **75 years**.
United Kingdom (abstraction licensing system in England and Wales)	A term of **12 years**, linked to cyclical reviews of water availability in a catchment, with the expectation of periodic renewal.

Figure 3.19. **Proportion of water allocation example specifying return flow obligations**

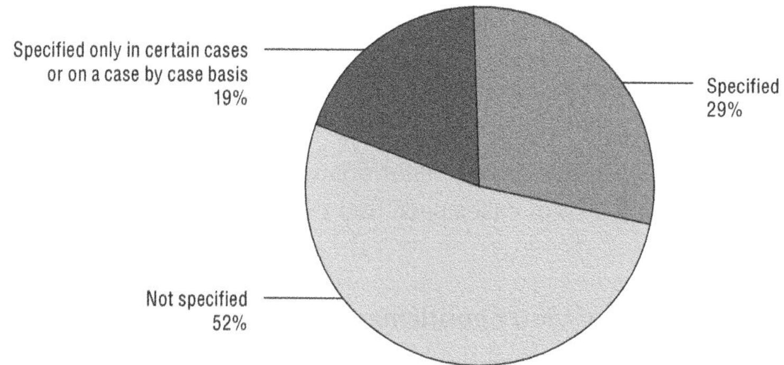

Note: Does not include: Prince Edward Island (Canada), Maipo River, 1st Section (Chile), Ubaté-Suárez River Basin (Colombia), the Dutch polder system in the western part of the country (the Netherlands), Switzerland, South Africa (Inkomati, Jan Dissels and Mhlatuse River Basins).

Source: See country profiles associated with this publication at *www.oecd.org/environment/water-resources-allocation-9789264229631-en.htm.*

Figure 3.20. **Consequences of non-use of water entitlements**
Number of examples

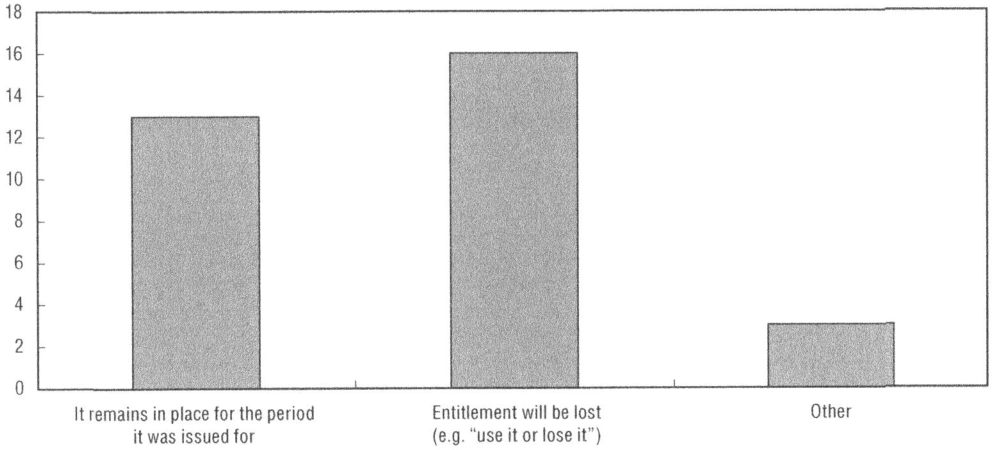

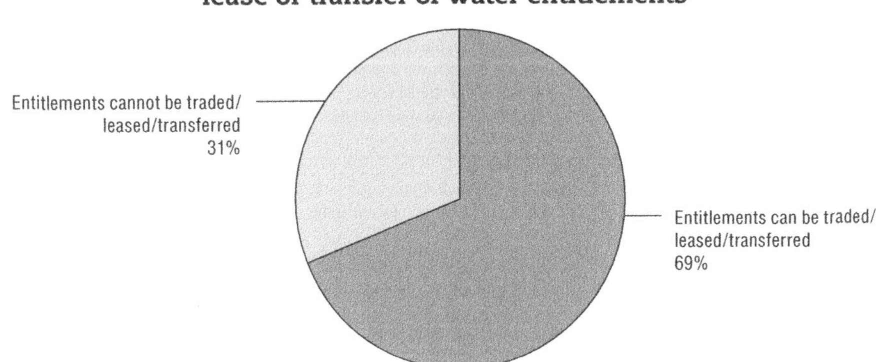

Note: Does not include: Newfoundland and Labrador (Canada), the Dutch polder system in the western part of the country (the Netherlands), Mexico, South Africa (Inkomati, Jan Dissels and Mhlatuse River Basins), and Switzerland.
Source: See country profiles associated with this publication at *www.oecd.org/environment/water-resources-allocation-9789264229631-en.htm.*

for, despite going unused. Alternatives to these two approaches were reported and typically reflected a modified version of one of these two approaches, by allowing the option to revoke entitlements after a certain lapse period of non-use or subjective to certain conditions.

Possibility to trade, lease or transfer water entitlements

To understand the flexibility given to water entitlement holders to make autonomous adjustments to allocation, respondents were asked to indicate if entitlements could be traded, leased or transferred in any way. As shown in Figure 3.21, in the majority of examples (69%), entitlements can be traded, leased, or transferred. Trading arrangements included more advanced, formalised markets to trade entitlements, such as the Murray-Darling Basin, Australia, Chile, Spain, or the abstraction licensing system in the United Kingdom as well as pilot projects on water entitlement trading, such as those being carried in some areas of the Yellow River Basin, China.

Figure 3.21. **Proportion of allocation examples that allow some form of trade, lease or transfer of water entitlements**

Note: Does not include: the Dutch polder system in the western part of the country (the Netherlands) and Switzerland.
Source: See country profiles associated with this publication at *www.oecd.org/environment/water-resources-allocation-9789264229631-en.htm.*

Allocation regimes reporting the possibility to trade, lease or transfer entitlements also indicated various (sometimes extensive) conditions on trade. These are summarised in Table 3.6. Only Israel reported that illegal practices, especially on farms, have become a serious issue affecting allocation and transfer of water resources.

Table 3.6. **Summary of various conditions on trade, lease or transfer of water entitlements**

Example	Summary of various conditions on trade, lease or trade of water entitlements
Australia (Murray-Darling Basin)	Trading occurs as a function of market demand and supply of water, which is in turn related to climatic and environmental conditions, commodity markets and the current and expected level of allocations. Prices reflect changes in demand and supply of water entitlements and contracts are exchanged between buyer and seller. Trades of entitlements (and lease or transfers) are subject to approval by licensing authorities (both the buyers and sellers, as these may be different).
Austria (surface and ground water systems)	In general, permits are linked to the location where the water abstraction takes place. Thus, a change of owners implies the transfer of the permit, but the competent authority has to be notified. There is no trading.
Brazil (São Francisco Basin and São Marcos Basin)	Water permits cannot be traded or leased, but can be transferred. Moreover, they can function as a financial instrument.
Canada (Alberta)	The transfer of allocation is voluntary, with a willing seller and buyer deciding for a price. There are no administrative charges associated with this trade. However, restrictions on trading are as follows: the transfer can be made only if the water can be physically transferred; the transfer must be authorised in a water management plan or through an order of Cabinet; the holder of the allocation must be in good standing; the transfer must not negatively impact other users or the aquatic environment. Furthermore, entitlements can function as a financial instrument.
Canada (Manitoba)	Transfers are permitted upon sale of the business (e.g. livestock operation, irrigation farm, water park, golf course, etc.). The new owner of the land must be able to put the water to beneficial use otherwise the Crown will not transfer the license.
Canada (Nova Scotia)	Only to be transferred through an assignment indenture.
Canada (Quebec)	All water withdrawal authorisations will be transferable. A transferee must, however, inform the ministry of the transfer within 30 days after it is made. This procedure will be in force when the new regime is adopted.
Canada (Yukon)	Can be leased or assigned to another user.
Chile (Limarí River Basin)	Transactions between owners of water rights can be made only under the authorisation of the General Direction for Water (DGA).
Chile (Maipo River, 1st Section)	Trading allocation is a function of the market demand and supply for water. No previous authorisation is required from any authority.
China (Yellow River Basin)	The Yellow River Conservancy Commission has not yet developed uniform water rights trading regulations, but some local governments are carrying out pilot projects. In 2003, pilot projects of water rights transfer were implemented in Ningxia and the Inner Mongolia Autonomous Region. Water-saving renovations in irrigation districts were funded by new industrial projects, reducing water losses and transferring the water saved to new industrial projects for payment. Thus far, 39 water transfer projects have been approved, with 337 million m^3 of water transferred. With no increase in total water consumption, newly-added water demand for socio-economic development is satisfied, thus promoting industrial restructuring and the transformation of economic development pattern.
Colombia (Ubaté-Suárez River Basin)	Water rights can be transferred totally or partially with prior authorisation. However, the competent environmental authority may deny it when reasons of public utility or social interest deem appropriate.
France [single collective management bodies for irrigation (OUGC)]	They can be transferred freely, for instance to the new land owner in some cases (Beauce area).
Israel (reuse of treated wastewater for agriculture, large scale desalination, municipal/regional water corporations)	Farmers can apply to the Ministry of Agriculture and request the reallocation of quotas. In some cases, reallocation is also done illegally.
Japan (Tone-Gawa River System)	Only if they succeed to obtain the approval of the river administrator.
Mexico	Rights can be legally transferred.
New Zealand (Waikato Region)	Transfer of surface water permits is provided in the Regional Plan subject to conditions (within the same catchment and water management class, for a use that already has consent or if permitted by the plan). Trading an entitlement requires a new permit or change to the permit and an assessment of the effect of the change. Transfers of groundwater permits work in a similar way. Trading is determined by individual arrangements between permit (entitlement) holders. Trading allocation requires a new permit or change to the permit and an assessment of the effect of the change. The Regional Council determines administrative costs for new permits or changes to permits.
Portugal (Tejo River Basin)	The user´s entitlement can be transferred through communication to the competent authority, with a minimum of 30 days in advance. They can also be traded or leased by notifying the competent authority, one month in advance. They can also be transferred to the heirs or legatees.
Slovenia	The right can be transferred to another person in the same manner that a new water permit is granted.
South Africa (Inkomati, Jan Dissels and Mhlatuse River Basins)	The existing user relinquishes the entitlement to use, on the condition that the new applicant gets the water entitlement. The beneficiary of a transfer is treated in the same way as new license applicant. Price is decided by the agreement between the buyer and seller. There is an administrative cost related to new applications. Water cannot be traded outside the basin.
United Kingdom (abstraction licensing system in England and Wales)	Buying or selling rights to abstract water require that an application is made and approval sought from The Environmental Agency to receive a new license or to vary an existing one. Restriction may apply.

Source: See country profiles associated with this publication at *www.oecd.org/environment/water-resources-allocation-9789264229631-en.htm.*

Water entitlements in some countries (Brazil, Canada, Chile, Peru, and the United Kingdom) can also function as a financial instrument to obtain credits, loans, capital or as requirements to start or add value to businesses. These examples are summarised in Table 3.7.

Table 3.7. **Summary of examples where water entitlements can be used as a financial instrument**

Example	Use of entitlement as a financial instrument
Brazil (São Francisco Basin and São Marcos Basin)	Water permits can function as a financial instrument. Banks usually require water permits from water users in order to concede loans. This procedure ensures that private parties will have adequate access to water when developing their enterprises.
Canada (Alberta)	Water entitlements can function as a financial instrument. The license in some cases can be used as collateral for obtaining credit or a loan from a financial institution.
Canada (Manitoba)	Water entitlements can function as a financial instrument, to be used at some financial institutions.
Chile (Maipo River, 1st Section)	Water entitlements can function as a financial instrument, for instance as a mortgage to get a credit/loan.
Peru (Parón River Sub-Basin)	Water licenses guarantee the development of public and private investment projects.
United Kingdom (abstraction licensing system in England and Wales)	Water entitlements can function as a financial instrument. A license can be sold and therefore has a financial value. It can also add value to a business that depends on water and therefore add value to the sale of a business.

Source: See country profiles associated with this publication at *www.oecd.org/environment/water-resources-allocation-9789264229631-en.htm.*

Pre-requisites to grant new water entitlements or expand existing ones

Nearly all of the allocation regime examples impose conditions on granting new water entitlements or expanding existing ones. Only two of the allocation examples surveyed indicated that there were no restrictions to granting such requests (Nova Scotia, Canada and Estonia). For the Murray-Darling Basin, Australia, the catchment is "closed" and access to surface water for new entrants is limited to purchasing entitlements (or leasing entitlements or purchasing allocations annually) from existing owners. Access to ground water is conditional on the assessment of third party impacts, environmental impact assessment (EIA) and existing users forgoing use.

Most allocation regimes reported at least one and often multiple conditions on new or expanding water use. Assessment of third party impacts is the most frequently cited conditions, with an EIA nearly as common. Existing user(s) foregoing use was also a frequently reported condition. Other conditions listed include the availability of unused water, adequate water balance, ecosystem improvement, equity, public inquiry, special authorisations, linking to planning instruments (summarised in Figure 3.22).

Users not required to hold water entitlements to access water

Most allocation regimes examples report that certain users or uses do not require formal water entitlements to access water. Typically, these exceptions are small scale uses relating to fulfilling basic human and animal needs, subsistence crop purposes or native people holding title rights. In many cases, examples clearly designate the mode of intercepting/ capturing water as by manual means only (collecting water in a bucket, for instance). These types of small scale uses are usually considered "insignificant" in terms of their impact of the overall resource. However, the actual volumes abstracted are not always well-documented or known to authorities, as the cost of monitoring such uses would greatly outweigh the

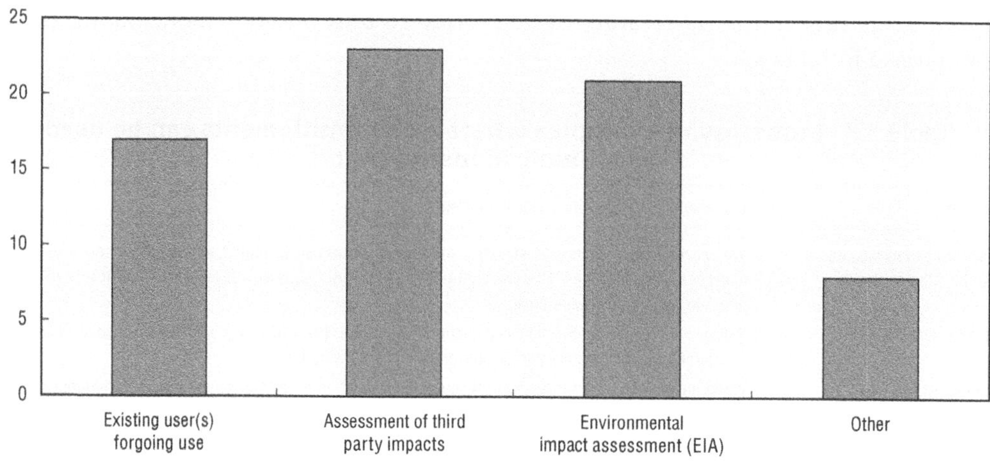

Figure 3.22. **Pre-requisites to grant new water entitlements or expand existing ones**

Note: Does not include: the Dutch polder system in the western part of the country (the Netherlands) and Mexico.
Source: See country profiles associated with this publication at *www.oecd.org/environment/water-resources-allocation-9789264229631-en.htm.*

benefits of formal control. Several allocation examples (Yukon Territory in Canada, the Yellow River Basin in China, Costa Rica, Luxembourg and the Waikato Region in New Zealand) also indicate water use for emergencies or to deal with exception circumstances or threats, such as firefighting, floods, droughts or other emergencies do not require entitlements.

Most allocation examples surveyed did not consider that these types of uses were becoming an issue, however the case of the Waikato Region, New Zealand is a good illustration of how these types of uses could pose a problem as their demand on the resource increases (Box 3.2).

Sequence of priority uses in water allocation

Nearly all allocation regimes surveyed have an established sequence of priority uses. In most cases, it is used to establish priority access to water during times of scarcity, when "exceptional circumstances" have been declared, such as in the case of drought. Some allocation regimes use the sequence of priority uses to determine which uses should receive water entitlements in cases where there is competition for access to water even in average conditions.

Most allocation regime examples define domestic and human needs as the highest priority use. Exceptions include the Netherlands, a small number of Canadian Provinces, water uses in Israel, and Peru. Several countries indicate several uses among the highest priority use. For instance, in the case of Austria, the highest priority use indicates both domestic (services for drinking water supply within sustainable abstraction limits) and the environment. In Brazil, both human and animal water consumption are designated among the highest priority uses. In the Waikato region of New Zealand, the highest priority uses include agriculture (milk cooling and dairy shed wash down), domestic uses, municipal supply, and renewable energy.

Box 3.2. **Unconsented water use in the Waikato Region, New Zealand**

In the Waikato Region, New Zealand, users not required to hold a water entitlement to abstract water include individuals taking water for reasonable domestic needs, or the reasonable needs of their animals for drinking water, and users taking water for firefighting purposes.* These uses are permitted so long they are not, or are not likely to, have an adverse effect on the environment. In addition, the Regional Plan also allows, *conditionally*, abstractions without a permit for supplementary surface and groundwater takes of a certain size (1.5-15 m^3/day) and temporary takes of 150 m^3/day for no more than five days per annum, and aquifer or well testing for 2 500/d for no more than 3 days.

Modelling was recently undertaken to estimate the magnitude of unconsented water use in the Waikato Region. Key findings from the study include:

- "Water use for dairy farming was found to have the most influence on model predictions. This was not surprising due to the high density of dairy cows in the region and the large volumes of drinking water required by lactating cows and the large volumes of water required for dairy shed operations.

- The relative water demand from permitted use and activities allowed to have unconsented water use in relation to the allocable flow was assessed in 202 catchments. In 35 of the catchments more than 50% of the allocable flow is taken for these activities alone, and in 16 of these the use exceeds the allocable flow. When consented authorised water takes are included with the permitted and unconsented water takes, there are 77 catchments with more than 50% of the allocable flow taken and of these, in 41 catchments the use exceeds the allocable flow.

- If intensification of dairying continues, the amount of animal drinking water required will for the most part increase without restrictions due to the high priority it is afforded by provision relating to unconsented water takes of the Resource Management Act. In many catchments this may result in nearly all the allocable flow being utilised solely for unconsented animal drinking water purposes."

Measures to address adverse impacts of an increase in these uses include monitoring permitted and unconsented takes and using these estimates to inform limit setting when flows and allocations are reviewed.

* According to national legislation [Sections 14(3)(b) of the Resource Management Act].
Source: www.waikatoregion.govt.nz/Services/Publications/Technical-Reports/A-Model-for-Assessing-the-Magnitude-of-Unconsented-Surface-Water-Use-in-the-Waikato-Region/.

A number of countries include water uses for national security purposes among the sequence of priority uses. In the Netherlands, for example, safety (preventing dyke collapse) and preventing irreversible damage to the environment are the highest priority use. In France, the cooling of nuclear power plants is considered a national security use. The most commonly reported second priority is either agriculture or environmental uses. Some allocation regimes have a very detailed designation of priority uses (6 distinct levels in the case of Hungary, Mexico, and Peru; 5 distinct levels in the case of Spain and South Africa). Others designate only one or two priority uses as compared to all others (São Francisco Basin, Brazil, Estonia, wastewater reuse in Israel, and Slovenia). The various ways in which the sequence of priority uses is defined across the allocation regimes surveyed are summarised in Figure 3.23.

Figure 3.23. **Sequence of priority uses in water allocation**

High priority ← → Low priority

Australia — The Murray-Darling Basin
1. Critical human water needs
2. Environment and transfer to the sea or another system
3. Agriculture, domestic and industrial

Austria — Surface and Ground Water Systems
1. Domestic (services for drinking water supply within sustainable abstraction limits) + environment
2. Agriculture; industrial; energy production; transfer to the sea or another system; national security (e.g. protection of infrastructure and critical dikes)

Brazil — São Francisco Basin
1. Human and animal water consumption
2. Others

Brazil — São Marcos Basin
1. Human and animal water consumption
2. Highly efficient irrigation
3. Hydropower production
4. Others

Canada — Alberta
1. Domestic
2. Agriculture traditional agricultural user
3. Industrial, energy production, environment prioritised by seniority

Canada — Quebec
1. Domestic
2. Environment
3. Agriculture, industrial, energy production and recreational

Canada — Manitoba
1. Human health and safety
2. Environment
3. Domestic
4. Agriculture
5. Industrial

Canada — Prince Edward Island
1. Fire control
2. Domestic
3. Environment
4. Agriculture, industrial, energy production

Canada — Newfoundland and Labrador
1. Environment
2. Domestic
3. Agriculture
4. Industrial
5. Energy production

Colombia — Ubaté-Suárez River Basin
1. Human community consumption (urban or rural)
2. Individual domestic needs
3. Farming community uses (aquaculture and fisheries)
4. Others (hydro power, industry, etc.)

Figure 3.23. **Sequence of priority uses in water allocation** (*cont.*)

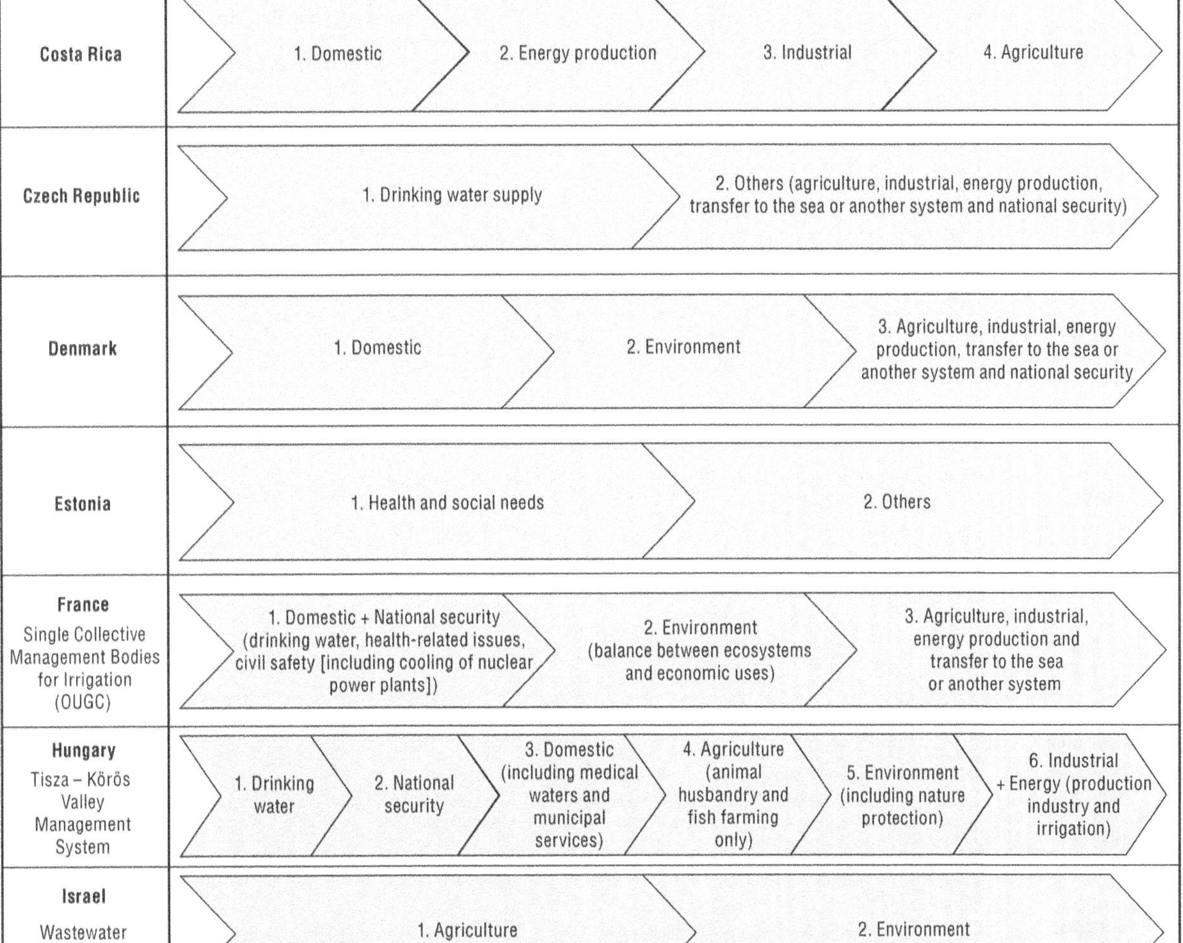

Figure 3.23. **Sequence of priority uses in water allocation** (*cont.*)

New Zealand Waikato Region	1. Domestic uses and municipal supply, agriculture (milk cooling and dairy shed wash down), hydroelectricity → 2. Industrial, energy production, environment, transfer to the sea or another system and national security (e.g. protection of infrastructure), agriculture (irrigation)
Peru Parón River's Sub-Basin	1. Environment → 2. National security → 3. Domestic → 4. Agriculture → 5. Energy production → 6. Industrial + transfer to the sea or other system
Portugal Tejo River Basin	1. Domestic → 2. Agriculture → 3. Industrial → 4. Energy production
Spain	1. Urban supply (incl. low levels for urban industries) → 2. Irrigation and agricultural uses → 3. Industrial uses for electrical energy production → 4. Other industrial uses → 5. Aquiculture, recreational uses and navigation and aquatic transport
Slovenia	1. Drinking water supply → 2. Others
South Africa Inkomati, Jan Dissels and Mhlatuse River Basins	1. Basic water needs and healthy aquatic ecosystems → 2. Meeting agreements with riparian countries → 3. Poverty eradication and greater equality → 4. National economy → 5. General economic purposes
Switzerland	1. Agriculture (cattle) and domestic uses → 2. Other priorities set by districts (cantons)

Source: See country profiles associated with this publication at *www.oecd.org/environment/water-resources-allocation-9789264229631-en.htm.*

Water abstraction charges

A majority of allocation regimes surveyed report that abstraction charges are in place. The proportion of allocation examples indicating that an abstraction charge is in place (breakdown by category of user) is summarised in Figure 3.24. Among categories of uses for which an abstraction charge is in place, industrial use is the most common. Nearly 70% of allocation regimes apply an abstraction charge to industrial users. Sixty-one per cent of allocation regimes apply a charge to agriculture, 58% to hydropower producers, 56% to domestic users, and 47% to energy production (other than hydropower). Examples of allocation regimes that do not have any abstraction charges include Austria,[8] Alberta and Prince Edward Island in Canada, the Limarí Basin and Maipo's River First Section in Chile, Denmark, and the Netherlands. In Spain, charges support cost recovery in the sense that the necessary infrastructure to meet water needs to charged to the users requesting such works. The two pricing instruments are the "regulation fee" and the "utilisation tariff".

Figure 3.24. **Proportion of water allocation examples with an abstraction charge**

Per category of use

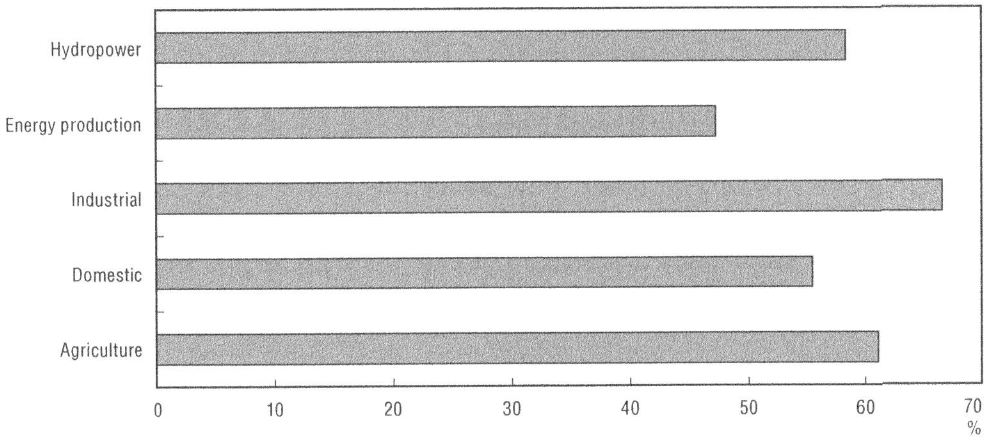

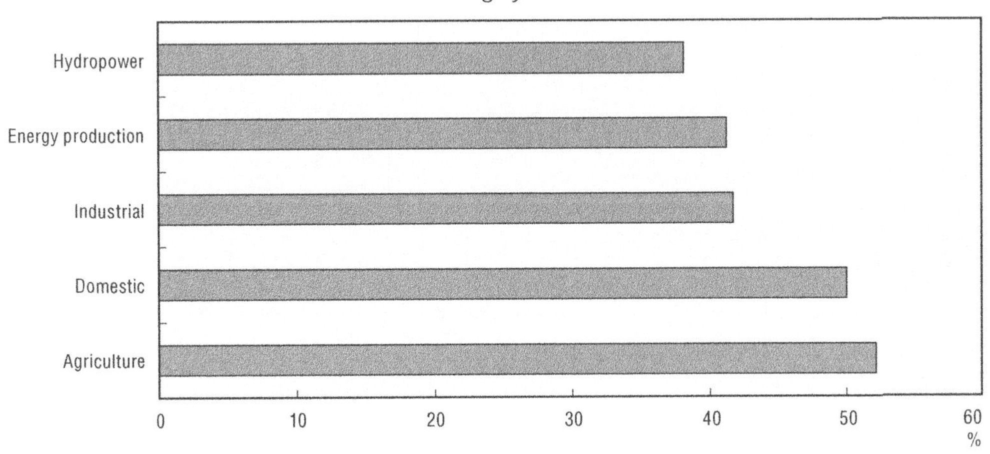

Note: Does not include Japan, where abstraction charges depend on each Prefecture.
Source: See country profiles associated with this publication at *www.oecd.org/environment/water-resources-allocation-9789264229631-en.htm.*

Among the allocation regimes with abstraction charges, volumetric usage is the most common basis for the charge. Seventy per cent of the allocation regimes that have an abstraction charge for industrial and domestic water use volumetric usage as the basis for the charge.

Of the allocation regimes that indicated that abstraction charges were in place, for most categories of use, fewer than half of the examples indicated that water scarcity is reflected (in one way or another) in the charge. In the case of agricultural use, for those allocation examples with abstraction charges in place, just over half indicated that scarcity is reflected in the charge (in one way or another) (Figure 3.25).

Figure 3.25. **Proportion of water allocation examples reflecting water scarcity in abstraction charge**

Per category of use

Note: Does not include Japan, where abstraction charges depend on each Prefecture.
Source: See country profiles associated with this publication at *www.oecd.org/environment/water-resources-allocation-9789264229631-en.htm.*

Data from the OECD/EEA *Database on instruments used for environment policy and natural resources management* provide a view of the ranges of levels of abstraction charges and their respective tax-base. Table 3.8 summarises examples of water abstraction charges among the higher and lower bounds of charges reported for select countries, based on available data.

Table 3.8. **Examples of water abstraction charges in select OECD countries**

Country/region	Examples of higher-end charges reported		Examples of lower-end charges reported	
	Specific tax-base	Tax rate	Specific tax-base	Tax rate
British Columbia, **Canada**	Industrial purpose – pulp mills	EUR 277 for 1 cubic foot per second	Commercial power use – output General power use – output up to a total of 160 000 mWh	EUR 0.7871 per MWh
	Oil field injection	EUR 246 per cubic foot per second[1]	Residential law or gardening watering (area exceed 0.25 acres)	EUR 7.6 per 10 acre feet a year
	Washing coal	EUR 239 per cubic foot per second	Mineral water sold in bottles or other containers	EUR 8.70 per 1 000 gallons a day or less
Estonia	Mineral water – drinking or for therapeutic baths	EUR 2.11 per m^3	Abstraction of surface water for cooling	EUR 0.0016-0.0072 per m^3
	Abstraction from Cambrium-Vendian groundwater aquifer for technological purposes (except food production)	EUR 0.15 per m^3	Water pumped out of open mines	EUR 0.017 per m^3
Germany (various *Länder*)	(Berlin) Charge on abstraction of groundwater	EUR 0.31 per m^3	(Bremen, Lower Saxony) Charge on abstraction of groundwater for fish farming purposes	EUR 0.0025-0.0026 per m^3
	(Schleswig-Holstein) Charge on abstraction of groundwater of users other than for public water supply	EUR 0.11 per m^3	(Bremen) Charge on abstraction of surface water bodies if the abstracted amount is more than 500 million m^3	EUR 0.003 m^3
Hungary	Effective rate on water abstraction of EUR 0.016-0.103 per m^3; varies by water source and region			
Poland	Groundwater abstraction for purposes other than production in cases where the water is in direct contact with food or medicine	EUR 0.026 per m^3 multiplied by differentiation coefficients depending on water quality and region	Surface abstraction for supply of households	EUR 0.0093 per m^3 multiplied by differentiation coefficients depending on water quality and region
Slovenia	The use of sand	EUR 2.45 per m^3	Abstractions for irrigation of agricultural land; breeding fish cyprinids	EUR 0.0008 per m^3
	The use of water for the operation of swimming	EUR 0.83 per m^3	Breeding salmonid fish species	EUR 0.0029 per m^3
United Kingdom	Abstraction charges are EUR 0.005 per m^3 on average			

1. Total not to exceed EUR 3 093.

Source: OECD/EEA, *Database on instruments used for environmental policy and natural resources management.*

Monitoring and enforcement of water withdrawals and allocation rules

Most allocation regimes report that they monitor water withdrawals and enforce allocation rules. In the case of Costa Rica, even though the Ministry of Environment and Energy has the legal authority to monitor withdrawals, it does not exercise control over withdrawals, due to a lack of human resources. Figure 3.26 depicts the proportion of allocation regimes monitoring water withdrawals, per category of use.

Figure 3.26. **Proportion of allocation regimes monitoring water withdrawals**
Per category of use

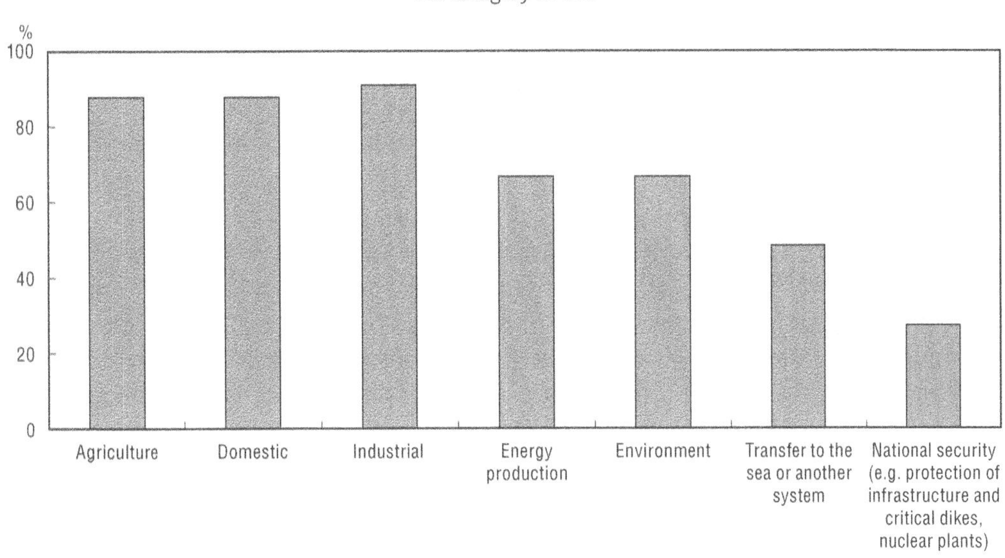

Note: Does not include: Denmark, Japan (Tone-Gawa River System) or South Africa (Inkomati, Jan Dissels and Mhlatuse River Basins).
Source: See country profiles associated with this publication at *www.oecd.org/environment/water-resources-allocation-9789264229631-en.htm.*

Industrial users are monitored in nearly all cases (91%) where monitoring occurs (except for the polder system in the western part of the Netherlands). Agriculture and domestic users are monitored in 88% of cases. Energy production and water for the environment are monitored to a lower extent (in 67% of cases). Transfers to the sea or another system as well as water for national security uses are monitored in 48% and 27% of cases, respectively.

Some type of sanction for transgression is in place in about two-thirds of allocation regimes surveyed. Sanctions are not in place in the following allocation regimes: Newfoundland and Labrador (Canada). Maipo River, 1st Section (Chile), Costa Rica, Tone-Gawa River System (Japan), Korea, Luxembourg, the polder system in the western part of the Netherlands, and in the Parón River's Sub-Basin (Peru). In cases where sanctions are applied, monetary fines are the most common type (reported in all regimes with sanctions, except for Portugal). Other types of sanctions include revocation of permits or other legal processes.

Summary of key findings from the Survey of Water Resources Allocation

The results of the survey provide a detailed view of the current allocation landscape. Findings indicate that most allocation regimes make use of elements of allocation design that encourage a robust system. However, the survey also reveals gaps in allocation regime design and identifies areas that could be adjusted to improve performance to reach the policy objectives of economic efficiency, environmental sustainability and social equity.

For example, at the system level, there are opportunities for improving the identification of the available resource pool to ensure hydrological integrity and a balanced repartition between *in situ* and diverted uses. Environmental flows are not secured in a number of allocation regimes. Systematically factoring in the potential impacts of climate change is also essential to ensure that allocation regimes can cope with shifting conditions. The design of sequence of priority uses has an important impact on how the risk of shortage is distributed across users and the benefits of water use that will most likely be foregone in times of scarcity.

At the user level, the robust design of water entitlements is critical to provide incentives for innovation and efficient use, investment, and efficient management of risk of shortage. All allocation regimes report having legally defined entitlements to access water granted either to individuals, to collective bodies or both. The duration of entitlements varies significantly, and a "use it or lose it" policy often applies. There is certainly scope to broaden the application of abstraction charges. Given that abstraction charges tend to be low in most cases, increases in charges would make low value and inefficient water uses less attractive. Also, many allocation regimes permit some form of trading, leasing or transferring entitlements, although a range of conditions placed on such activities was reported. Table 3.9 captures the summary of main findings of the survey.

Table 3.9. **Summary of main findings of the Survey of Water Resources Allocation**

Elements of an allocation regime	Main findings from the survey
	System level elements
Clear legal definition of the ownership of water resources	● The large majority of countries indicate that water resources are publicly owned (or designated as "ownerless property"). Nearly all instances of privately owned water resources relate to ground water, which is owned by the owner of the land on which it resides. ● There can be ambiguity between various legal regimes within a given jurisdiction (e.g. customary rights versus rights designated in different laws; see for example, Japan or Korea). This legal "pluralism" is a source of conflict among water users and increases the likelihood of litigation.
Institutional arrangements for allocation	● Slightly fewer than half (48%) of countries indicated a role for the Ministry of Environment in water allocation. Even fewer (30%) indicated a role for a basin authority.
Identification of available water resources	● A majority (73%) of allocation examples reported that they are neither over-allocated nor over-used, with 11% considered over-allocated, 11% over-used. In 2 countries (the Maipo River, 1st Section, Chile and various regions in Spain) the water resource is considered both over-allocated and over-used. ● Most water systems in the examples surveyed are regulated to some extent (either partially or fully) so that there is some control over the flow rate. ● Half of the examples surveyed reported agriculture as the dominant water user.
Identification of in situ requirements/definition of available ("allocable") resource pool	● Three quarters of countries report that environmental flows are defined, but there are considerable variations in the methodologies used to do so. ● Only 57% of allocation regimes report taking into account climate change, in the definition of the available resource pool.
Abstraction limit ("cap")	● While a significant majority of allocation regimes (92%) have a clear definition on the limit on consumptive use, few rely on flexible limits (defined in terms of the proportion of the resource that can abstracted, instead of a fixed volume.
Definition of permitted uses not required to hold an entitlement	● Most allocation regimes allow certain users or water uses to access water without formal water entitlements. These exceptions are usually small scale uses relating to fulfilling basic human and animal needs, subsistence crop purposes or native people holding title rights. They usually do not pose an issue for the sustainable management of the resource, however there are some notable exceptions (see, for example, New Zealand).
Definition of "exceptional circumstances"/sequence of priority uses	● A sequence of priority uses is clearly established in nearly all allocation regimes. Most regimes define domestic and human needs as the highest priority use.
Requirements for new entrants	● Nearly all regimes impose conditions on granting new water entitlements or expanding existing ones. An assessment of third party impacts and an environmental impact assessment are the most frequently cited conditions.
Mechanisms for monitoring and enforcement	● Most allocation regimes (except Costa Rica) report that they monitor water withdrawals and enforce allocation rules. Industrial users are the most frequently monitored (91%) with agriculture and domestic users monitored in 88% of cases. ● Two-thirds of regimes report that sanctions are in place for non-compliance with the rules and regulations of allocation regimes. Monetary fines are the most common type.

Table 3.9. **Summary of main findings of the Survey of Water Resources Allocation** (cont.)

Elements of an allocation regime	Main findings from the survey
	User level elements
Clear, legal definition of water entitlements	• Water users' entitlements are legally defined in all allocation regimes, with the exception of the Netherlands. The majority (88%) allow for private entitlements. Regimes that allow entitlements to be granted to either an individual or a collective organisation (e.g. water users association, municipality) were more common than those that allow for only individual entitlements.
Abstraction charges	• A majority of regimes charge for water abstraction. Industrial use is the most common type of use to have an abstraction charge (nearly 70% of regimes). Volumetric charges are the most common basis for the charge.
Specification of return flow obligations in water entitlements	• Around half (52%) of allocation regimes do not specify return flow obligations of water entitlements.
Duration of water entitlements with expectations for renewal	• In most cases, water entitlements are time bound, either with or without an expectation of renewal. In a few cases are water entitlements granted in perpetuity (Australia, Chile, Israel, and Peru), with or without requirements for beneficial use or continuity of use. • Slightly more allocation regimes reported using a "use it or lose it" system for un-used entitlements than regimes reporting that entitlements remain in place for the period they are issued for, despite going unused.
Possibility to trade, lease or transfer water entitlements	• Two-thirds of allocation regimes allow for some sort of trade, lease or transfer of water entitlements. Specific conditions to trade, lease or transfer usually apply and often require the review and approval of an authority.

Source: See country profiles associated with this publication at *www.oecd.org/environment/water-resources-allocation-9789264229631-en.htm.*

Overall, the survey provides a solid basis on which to identify the policy levers available to improve the performance of allocation regimes. This empirical basis has been used to inform the development of policy guidance on allocation regimes in Chapter 5.

Notes

1. The questionnaire used for this survey has been included in the Annex A.

2. This analysis is based on country profiles prepared by the OECD Secretariat, based on the responses collected via the questionnaire. The country profiles are available on a dedicated website.

3. With the exception of Canada, which provided this information at the provincial/territorial level.

4. It is noted where certain examples are not included in the analysis because an answer was either not applicable or not provided.

5. The programmes include: purchasing entitlements from willing sellers at market prices and implementing infrastructure works (both on farm and in irrigation delivery systems) to reduce losses and improve water efficiency. State governments also undertake similar efforts.

6. It should be noted that some responses provided information with a different metric, depending on the information available. Detailed information for each specific allocation regime is reflected in the country profiles.

7. Although renewal is not expected for water consents granted in the Waikato Region (New Zealand), factors such as prior investment or existing infrastructure would be considered and may influence decisions about a request for renewal.

8. While users do not pay abstraction charges separately, users are required to pay the full cost of water services.

References

OECD (2015), *Survey of Water Resources Allocation: Country Profiles*, *www.oecd.org/environment/water-resources-allocation-9789264229631-en.htm.*

OECD/EEA, *Database on instruments used for environment policy and natural resources management*, *www2.oecd.org/ecoinst/queries/* (accessed 17 April 2014).

Chapter 4

Reforming water allocation regimes

This chapter examines common themes related to reform, drawing on case studies documenting the experience of water allocation reform in 10 OECD and BRIICS countries. Although water allocation reform is inherently a political process unique to its time and place, valuable insights can be drawn from the experience of other countries. These can be instructive for those contemplating allocation reform or actively pursuing it. This chapter examines the drivers of reform, the process of identifying and selecting reform options, stakeholder engagement, and other key aspects of the reform process, drawing out insights and lessons learned.

Key messages

- Engaging in an **appropriate policy dialogue to support a water allocation reform** can help to avoid adopting an overly technical and technocratic approach to reform. The *"Health Check" for Water Resources Allocation* (Chapter 5) can provide useful guidance for such a dialogue.

- Water allocation reform is not a discrete, time bound process. It tends to be an iterative process, which **extends over many years or even decades**. Institutional path dependency can raise the cost of improving the flexibility of allocation to respond to changing or novel conditions.

- **Concerns about water scarcity** and **insufficient water for ecosystems** are often cited drivers of allocation reform, along with broader political or structural reforms. Droughts can provide a salient, visible event to trigger action.

- The reform process allows for **ample opportunities for participation and negotiation**. **Willingness to engage stakeholders** and appropriately **compensate** potential "losers" facilitates the process. Compensation can take various forms, such as financial transfers or permission to build storage structures.

- Before introducing changes to an allocation regime, it is essential to determine a **sustainable baseline** (how much water is available for allocation once *in situ* requirements, including for the environment, have been satisfied) and consider possible **unintended consequences**.

A significant number of countries are either currently reforming their allocation arrangements or have recently done so. Findings from the OECD Survey on Water Resources Allocation indicate that 60% of countries surveyed have recently reformed their allocation regimes. Over half of respondents indicated that allocation reforms are ongoing. In this analysis, "reform" is understood broadly, to encompass both transformational reforms that can entail fundamental changes to significant aspects of an allocation regime, as well as incremental changes in the policies, laws, and mechanisms that have a tangible impact on allocation arrangements.

Reforming water allocation arrangements can be a very difficult political challenge. Reforms can lower the volume of water some users will have access to, they can change the distribution of the risk of shortage across water users, and they can affect infrastructure and investment needs. Even incremental changes to an existing allocation regime can create opposition and require costly compensation to free up water by buying out existing water users.

Although water allocation reform is inherently a political process unique to its time and place, valuable insights can be drawn from the experience of other countries in dealing with political economy challenges related to reform. These insights can be instructive for those contemplating allocation reform or actively pursuing it. Drawing on case studies documenting the experience of water allocation reform of 10 OECD and BRIICS countries, this chapter draws out useful insights and lessons learned. It examines common themes related to drivers of reform, the process of identifying and selecting reform options, the reform process itself, and assessments of reforms that seek to determine if they reached their aims.

Why reform? Building the case for water allocation reform

It can be difficult to pinpoint the key drivers of allocation reform. Often the reform is driven by an accumulation of factors over time. The impetus from a "trigger event" can also help to spur reforms on. Nevertheless, looking across a range of country experiences, some common themes emerge. Concerns about water scarcity and insufficient water for ecosystems due to growing demand and/or reduced supply have provided the rationale for reviewing existing allocation arrangements in a number of countries. The case of Alberta, Canada illustrates the need to adjust to changing conditions, such as increasing demand, by maturing the existing allocation regime which had been previously fit for purpose, but was no longer seen to be so, given changing circumstances. In some cases, authorities have had to claw back entitlements to over-allocated and over-used resources, or find alternative sources of supply, such as in the case of Israel.

A drought is often seen as a good opportunity to make the case for change, even if change is long overdue. Such a crisis can help to catalyse action, as authorities are under pressure to respond to both the short term, acute crisis as well as engage in a longer term process, aimed at preventing future crises. Motives outside of the water domain, such as broader political reforms to improve economic efficiency (e.g. Australia, Israel) or equity in the distribution of resources (e.g. Chile, South Africa) have driven allocation reform. However, as illustrated by the case of England and Wales, a well-documented and broadly accepted case for reform does not necessarily translate into immediate action. Finally, two cases clearly highlight that perceived risks to the government, such of the threat of litigation, can either spur reform (as in the case of France) or, on the contrary, delay its implementation (as in the case of South Africa).

Allocation reforms have also taken place in the broader context of a general shift in water management approaches over the past several decades. Until the 1980s, water resources management in most OECD countries put the emphasis on infrastructure, or "supply-side" technical solutions, to aim to harvest the maximum amount from the resource. This "supply side" approach has since shifted to place greater emphasis on the sustainable management of the resource with reliance on "demand side" economic solutions, including a greater role for market-based allocation mechanisms (see, for example, OECD, 2010).

As illustrated in several cases, there are multiple motives that drive allocation reform and these are dynamic over time. The focus of policy objectives for allocation can also shift over time. A common pattern is a progression from a focus on regional development, towards more efficient water resource use, to increased attention to the value of water for sustaining ecosystem services. Equity concerns arise in multiple stages, often in an attempt to remedy past allocations that are no longer perceived as fair, or to address unintended consequences of previous reforms. This section will examine each of these reform drivers in turn.

Concerns about growing water scarcity and lack of water for the environment

Concerns about growing water scarcity and lack of water for the environment is a frequently cited reason for reviewing existing allocation arrangements. Concerns about scarcity can be fuelled by increasing demand or shifts in water availability, or due to degraded quality, or climatic shifts. In particular, droughts can provide a salient, visible event to trigger action (see Box 4.1). Efforts to secure adequate water for ecosystems can be driven by new legislation, such as the EU Water Framework Directive, or visible environmental deterioration, such as rivers running dry. In some cases, reforms have sought to address issues that have arisen from previous allocation reforms. The cases of Alberta, Canada, Australia, Chile, Israel, England and Wales, and South Africa provide insight on this theme.

In the case of Alberta, Canada, concerns about meeting future water requirements arose with increasing urban development (notably in the southern half of the province) and a drought in 2001-02. By the early 1990s, there were already conflicts between farmers and environmentalists over low flows in the rivers. At the time, there was little experience of monitoring actual water use and managing water shortages. The situation prompted a review and adjustment of allocation arrangements, which the provincial government characterises as a process of maturing rather than reforming, *per se*. The old system worked well given the prevailing circumstances, but the maturing of the system became inevitable as water demand exceeded available resources. A review of the Water Resources Act of 1931 and past policies was undertaken. The process took five years and resulted in the enactment of the Water Act of 1999. The concerns regarding growing scarcity resulted in authorities halting the issuance of new water licences in certain basins and establishing a mechanism that allowed existing licenses to be transferred among users.

In a similar vein, Israel faced rapid change in demographic and economic trends that put water resources under significant pressure. Hence, the increasing shortage of natural water resources was the major driver for the reform. Over the last 18 years, the average renewable quantity of water dropped to an average of about 1.2 million cubic meters per year (MCM/Y) from an average of about 1.4 MCM/Y over the last 50 years (Feinerman et al., 2013). The water sector was under constant pressure also as a result of a substantial increase in population

> ## Box 4.1. **Never let a good drought go to waste**
>
> Winston Churchill's oft-repeated quote "never let a good crisis go to waste" alludes to the opportunity that such events create to change the *status quo*. This also applies to water resources reform, where droughts provide a salient event that makes visible the impacts and costs of water scarcity and poorly managed allocation. Even when a case for allocation reform has been long recognised, a drought can provide the needed trigger to spur action on the reform agenda. This is reflected in several cases. For instance, in New Mexico, United States, the need for reforming the water allocation regime had been built up during a period of several decades. The very severe drought in 2002 can be seen as a catalytic event that advanced reform. The entire state was considered a drought disaster area and all users suffered from shortage. Further, the state had difficulties complying with its obligation of delivering water to the neighbouring state Texas, as agreed under the Pecos River Compact (Bossert, 2013). In California, United States, a multi-year drought in the late 1980s and early 1990s helped the water market take off by spurring direct state purchases and the development of an emergency drought water bank. In the case of England and Wales, coincidentally, the *White Paper* "Water for Life" which set out the proposals for abstraction reform was published during a severe drought, while culminated in a risk of supply failure to London in 2012. This raised expectations about the abstraction licensing reform and a major stakeholder engagement process was started. Finally, the heat wave of 2003 in France led to the development of the National Plan against droughts (*Plan sécheresse*) in 2004, which laid some of the groundwork for the subsequent allocation reform.

(multiplied almost twelve-fold over the past 60 years) and an increase in the standard of living. Up until a few years ago, water supply was almost entirely dependent on the renewable-storage capacity of three main natural sources (Feitelson, 2013).

However, the water crisis also resulted from previous policy decisions, which resulted in over-allocated resources. During the first decades of the Israel's existence, water allocation policy gave priority to accelerated economic development, particularly in the agricultural sector, over the naturally available quantities. This caused a continuous and increasing erosion of the operational storage capacity in the natural sources (which worsened during drought years) up to a "crisis" when shortage amounted to almost equal the level of annual overall consumption. This has occurred twice since 2000 (State Comptroller, 1990; Bain Committee, 2010).

For South Africa, there was a growing recognition of increasing water scarcity and pollution challenges arising from industrial expansion and population growth, along with the need, in line with the concept of sustainable development, to protect the aquatic ecological base. South Africa is a water scarce country, with large semi-arid areas. In addition, extensive mining has left a legacy of water pollution, exacerbated by industrial and agricultural practices. An intricate network of infrastructure transports water between catchments to areas of high demand. However, the options for further infrastructure development were increasingly expensive and alternative options to expand water supply were needed. Several of South Africa's most economically important river basins were facing closure and increasingly complex approaches were required to reconcile water availability and demand. A long history of managing scarce and variable water resources established a core of experienced water managers who foresaw that greater control over and understanding of the water allocation process held the key to maintaining economic growth in the face of growing scarcity. There was also a strong sense that the existing legislation was based on the legislation of well-watered European countries and needed to be amended to address the water issues in a water scarce country with high levels of climate variability. Moreover, the introduction of legislative requirements for environmental flows introduced new challenges, potentially requiring the State to claw back allocated water in some places. This required a shift away from the riparian doctrine, in which land owners are entitled to a fair share of water on or under their properties, to an allocation system based on usufruct, quantified entitlements.

England and Wales faced similar challenges, but years passed before there was sufficient momentum for moving the reform forward. Strains on the environment due to pressures on water resources had been long recognised. The growing recognition of the scarcity of resources (counter-intuitive to the perception of England as a wet country) and the unacceptability of dry rivers, led to calls for reform in the early 1990's. In 1997 the then Deputy Prime Minister gave a government commitment at a Water Summit (called after a very severe drought in 1995-96) that the abstraction licensing system would be reviewed. Despite ongoing public pressure for reform to progress, legislation in 2003 merely delivered some minor amendments to the then 40-year old system. In 2011 the momentum was regained when the government published the *White Paper* "Water for Life", which included firm proposals for abstraction reform. Crucially, it was supported by "The Case for Change", which was an assessment by the Environment Agency of the potential impact on water availability of climate change, coupled with increased demand for water from population growth and energy supply (Box 4.2).

Like England and Wales, escalating environmental concerns were among the key drivers of allocation reform in Australia, along with broader reforms to promote economic efficiency. From 1981 to 1983, the River Murray mouth closed for the first time since regulation of the river system began, leading to increased awareness of environmental water requirements. Over time, as the water market developed and water trading expanded, it became clear that not only was the system over-allocated, but the cost of not dealing with the issue would increase severely in the future. As a response, the Australian government introduced in 2007 the Buyback programme to purchase water entitlements for the Environmental Water Holder from voluntary sellers in the market.

Similar to Australia, Chile also faced challenges related to the over-allocation of water resources in a market-based system leaving insufficient water for the environment. Recent reforms in 2005 included amendments to the prevailing allocation regime that sought to correct issues related to social equity and environmental sustainability that were not reflected in the Water Code of 1981. Prior to the 1990s environmental and water management policies did not pay much attention to meeting water requirements of the environment and as a result, the totality of river flows was allocated to users. This gradually changed. In 1994, an improvement was made in the system of Environmental Impact Assessment that proposed the establishment of an ecological flow as a mitigation measure against potential negative environmental externalities associated with water trading. Amendments to the Water Code in 2005 formally established minimum ecological flows to be considered in the granting of new water use rights by the Directorate General of Water (DGA).

Box 4.2. **Making the case for water abstraction reform in England and Wales**

The example of England and Wales demonstrates the value of making a clear case for change, including ensuring that the shortcomings of the current system are widely understood. To support the proposals for abstraction reform set out in the *White Paper* "Water for Life" (2011), the United Kingdom Environment Agency developed "The Case for Change", which assessed the potential impacts on water availability of climate change, coupled with increased demand for water from population growth and energy supply needs. The assessment considered 11 emission scenarios from the United Kingdom Climate Projections (UKCP09) and their potential impact on river flows.

The conclusions were varied, but suggested that average summer river flows could reduce by up to 80%. This reduction would not be offset by increased winter precipitation. Attempting to maintain current (aspirational) levels of environmental protection would reduce significantly the water available for abstraction. Similarly, demand pressures under a range of socio-economic scenarios would pose a threat to environmental limits. The analysis considered averages; drought events would have greater impact. The south and east of England is already water stressed from a combination of low rainfall and high demand. But the wetter west and north could not be relied upon to make up any shortfall: the analysis suggested that these catchments would be most affected by climate change consequences.

The "Case for Change" spelt out the future challenges and pressures on water resources. However, in order to understand how the allocation system should be reformed to respond to those challenges it was important to understand its shortcomings. In December 2013 the government published "Making the Most of Every Drop", a consultation on abstraction reform, which set out why reform was necessary:

- The current system does not systematically link access to water to availability. Only a quarter of licences have controls to stop abstraction to protect the environment or other abstractors during periods of low availability. Conversely, the system struggles to allow additional water to be taken during higher flows.

- The system does not help abstractors to trade available water effectively, and to provide price signals to promote efficient water management.

- Abstractors are not currently incentivised to manage water efficiently.

- Much of the water (generally less than half) that is licensed is not actually used. This potentially denies access to others.

- The current process to change most licences that are causing damage to the environment is expensive and time consuming. As the climate changes and flows reduce or become more variable, more licences are likely to require changes, making this problem much worse and more expensive.

- The system fails to incentivise abstractors to manage risks from future pressures on water resources.

The government stated in its consultation that "these weaknesses may constrain economic growth due to reduced resilience and getting sub-optimal economic value from available water, while not efficiently protecting the environment". Officials from the Department for Environment, Food and Rural Affairs (Defra) worked closely with the Environment Agency, and representatives from a wide range of other organisations, in order to fully develop the policy options for reform.

Source: Barker, I. (2014).

Overall, these examples highlight how concerns about growing water scarcity and insufficient water for ecosystems can figure among the key allocation reform drivers. These strains can arise from rapid demographic and economic changes, droughts, and climate change, or over-allocation of the resources for consumptive uses. Although rivers running dry can provide a powerful signal to policy makers and society, concerns about the environment can be broadly recognised for a long period of time, without necessarily spurring action on their own.

Motives from outside of the water domain

Beyond concerns about the state of water resources, motives from outside of the water domain are often a contributing, and even decisive, factor in allocation reform. Two factors are often cited: the drive to improve economic efficiency of resource use and the equity of resource use. Perceived legal risk can also spur or hinder reforms, depending on the context (Box 4.3).

Box 4.3. **Perceived legal risk: A help or hindrance for water allocation reform?**

In addition to the reform drivers discussed in this section, the cases of France and South Africa highlight the influence of perceived risks by authorities on allocation reform. Interestingly, the perceived risk has opposing effects depending on the specific example. In France, rising judicial risk spurred action to deal with a depleted aquifer. In another instance, proposed revisions of licenses were stalled after water users took the authority to court. In South Africa, the perceived risk of potential litigation had stalled the implementation of allocation reform. The example from South Africa is further discussed in Box 4.5.

In France, when water tables ran dry in one particular aquifer (nappe de Beauce), environmental NGOs took notice and action. Faced with a rising environmental risk (frequent and damaging low flows), the government created special zones for water sharing (*zones de répartition des eaux* or water apportionment areas, WAAs, in English) in 1994. WAAs are areas that have known regular water stress situations, where access to water can be restricted (stricter conditions to get abstraction licenses and greater water abstraction taxes). WAAs are in accordance with the 1992 Water Law. They have been revised in 2003.

In 2003, France was affected by a heat wave, which led to a national plan against droughts (*Plan sécheresse*) in 2004. Before the plan passed, some local initiatives had been taken to reform allocation regimes. For instance in the Poitou-Charentes region, additional demands for water license triggered tensions between water users and NGOs. The local State representative ("Prefet") revised the water licenses, with a view to address over-allocation. Water users took the State to court, and the State lost all the cases because the decision was not backed by a legal procedure: granted licenses could not be withdrawn that easily. Changes had to be anticipated and compensation had to be negotiated according to the local situation. This particular experience confirmed that the State was facing a judicial risk, if it was to revise existing water licenses. Facing rising judicial risks, France considered that the *status quo* was not an option, and hence examined options for allocation reform.

The drive to improve economic efficiency was among the key reform drivers for Australia, Chile, and Israel. For Australia, in addition to the escalating environmental concerns discussed above, the pursuit of economic efficiency was a decisive factor behind the establishment of a comprehensive water market in the Murray-Darling Basin.

Similar to Australia, reforms in Israel were linked to broader economic reforms. The Ministry of Finance established the need to continue *"structural reforms to further enhance the public sector, increase competition, reduce the public expanse and increase its efficiency, and promote a responsible and growth-supportive fiscal policy..."* (Government of Israel, 2013). In this framework, aspects of the water sector and its management were extensively addressed. From the early 1990's a number of reports by the State Comptroller and several inquiry committees pointed at administrative and structural failures regarding a multitude of aspects in the water sector policies. It is mainly the resulting recommendations and their implementation which led the reform (Arlozorov Committee, 1997; Magen Committee; 2002; Bain Committee, 2010).

For Chile, the economic liberalisation of 1973-89 was a key driver of water reforms in the 1980s. The economic paradigm had shifted from a state-controlled economy to an emphasis on market-based approaches. Although private water use rights existed in Chile prior to the Water Code of 1981, the legislation at the time restricted the creation and operation of efficient water markets consistent with the new economic system. These restrictions related mainly to ways in which water use rights were defined, the amount of information available to users, transaction costs, the way third-party impacts were dealt with, the approach to conflict resolution, and the legal framework necessary for the market to operate properly (Donoso Harris, 2003; Donoso Harris, 2011). The enactment of the Water Code of 1981 sought to remove these constraints and created legally defined tradable water use rights to facilitate economically beneficial reallocation of water (Bauer, 2004; Büchi, 1993; Hearne and Donoso Harris, 2005).

Subsequent reforms in Chile sought to remedy some of the issues related to equity and environmental sustainability that arose from earlier reforms. For example, with the introduction of Water Code in 1981, a rapid increase of requests for water use rights for speculation and hoarding purposes occurred,[1] which resulted in monopolistic behaviour and a reduction of water resources available for allocation to other potential uses (even though they were not actually being used). This created an impediment to the development of new investment projects on account of not being allowed to acquiring new rights (Pena et al., 2004; Bitran and Saez, 1994; Donoso Harris, 2003; Donoso Harris, 2011).

To address this speculation and hoarding, the government introduced a non-use tariff for unused water. Once the water use right is determined to be "unused", the tariff is levied, based on a system of escalating charges.[2] As a result of the reform of the Water Code in 2005, along with other measures, speculation and the hoarding of non-consumptive water use rights have been reduced (Pena, 2010) freeing up water to be accessed by a broader number of potential users, thereby improving the equity of allocation.

Equity in allocation was a primary driver of water reform in South Africa. Water allocation reform in South Africa was driven during the political transition to democracy in 1994 and formed part of a broader suite of legislative reforms aimed at fundamentally transforming the South African political and economic context. The primary driver was the need to transform a society in which the black majority had been excluded from access to natural resources (including water) or the benefits derived from such natural resources.

Thus, in 1994, around 95% of the water used in South Africa was in the hands of the white minority. The water allocation reform proposed in the 1997 *White Paper on a National Water Policy for South Africa*, and the ensuing National Water Act were aimed at addressing this historical injustice.

Policy options appraisal for water allocation reform

The appraisal of policy options is a key step in the reform process. Focussing on remedying the perceived shortcomings of the existing system strongly influenced the identification of options in several countries. The cases studies also show that international experience is frequently considered in the course of developing policy options. However, different countries may draw different lessons from the same experience. At the same time, policy options can be constrained by limitations imposed by other policy areas (e.g. tax policy), strong opposition by stakeholders, broader political or economic considerations, or conflicts with existing well-entrenched principles (e.g. first in time, first in line). Finally, while it can be difficult to quantify all of the costs and benefits of alternatives, economic assessment has proven to be useful in option selection. This section explores the process of developing policy options across these themes, with illustrations from the case studies.

The perceived shortcomings of the existing water allocation system strongly shapes the reform options considered

Focussing on remedying the perceived shortcomings of the existing system significantly influenced the identification of options in several countries. The cases of Alberta, Canada, England and Wales, France, and South Africa provide useful illustrations. In addition, a few cases demonstrate how allocation reform options can be differentiated within a country, to address varying degrees of scarcity, or risk of shortage (Box 4.4).

In Alberta, Canada, the concerns regarding growing scarcity influenced the development of policy options to improve the allocation regime. As a response to growing scarcity, authorities halted the issuance of new water licences in certain basins and established a mechanism that allowed existing licenses to be transferred among users. New rules were created with provisions for permanent and temporary (in-season) transfers and trades, subject to the approval of the Alberta Ministry of Environment and Sustainable Resource Development (Alberta Environment, 2003).

Changes in allocation arrangement were also made to address the issue of licenses held for water that was not being put to beneficial use. To address this, authorities introduced a new feature for all new water licenses issues after the Water Act took effect. All new water licenses would have expiry dates and the licence holders would have to apply for renewal. This enabled reallocation of water that has not been put to beneficial use. However, this change met with some concern, as imposing expiry dates could be interpreted as taking away licensees' property rights.

Similar to Alberta, in England and Wales addressing the shortcomings of the current system played a key role in working up and testing options for consultation. In "Water for Life" the government substantially revised the 1999 policy objectives, as follows:

- to send clear signals on water availability, so as to drive adaptive behaviour by abstractors

- to better reflect the value of water to customers and ecosystems

---

Box 4.4. Differentiated approaches tailored to different degrees of water scarcity

A differentiated approach to allocation can be used to tailor reform options to different circumstances. Both the cases of South Africa and England and Wales provide good illustrations.

South Africa's allocation reform recognised different approaches should be taken depending on water scarcity and the potential economic impact of allocation reform in a particular area, as follows:

- In catchments where **water availability was not likely to limit growth**: the focus would be on actively seeking opportunities for viable black or women owned enterprises who could be allocated water entitlements, without compromising existing users, or relicensing users.

- In catchments where **water is becoming limited, but which were not prioritised for compulsory licensing**: the process would not actively seek new water users, but would encourage water trading among water users. If viable black or women owned enterprises made application for licences, these would be issued, in effect slightly lowering the assurance of supply to existing users. Should the volume of water required by new Broad-Based Black Economic Empowerment users warrant it, the catchment would be prioritised for compulsory licensing.

- **Catchments prioritised for compulsory licensing:** in these catchments, a rigorous process of assessing the availability of water for allocation, determining the extent of existing lawful water use, determining the opportunities for uptake by black and female users, and assessing the impacts on existing users would be followed.

For England and Wales, the reform components to better link abstraction to flows and facilitate trading would only be introduced in catchments where there were clear environmental and economic benefits due to water scarcity and the potential for trading – so-called "**enhanced catchments**", where much of the benefits of reform would be found. Catchments that do not show clear environmental or economic benefits for enhanced reform would undergo basic reform only – "**basic catchments**". However, as the climate changes and the demand for water increases, the number of basic catchments is likely to decrease. The impact assessment of the reform took account of the way in which particular elements of reform were likely to be implemented. Both reform options were estimated to have set up costs of GBP 10-16 million, with water shares being more expensive because of its increased complexity.

Source: Quibell, G. (2014); Barker, I. (2014).

- to recognise the value of discharges
- to drive efficiency in water use
- to be fair and equitable
- to be flexible and responsive (this was seen as one of the key aims in order to help manage the uncertainties inherent in climate change and demand forecasts)
- to meet water needs for people and the environment at least cost.

These policy objectives were an effective litmus test of the options appraisal, set against the assessment of the shortcomings of the current system. The first option, "Current System Plus", would build upon existing good practice to strengthen the link between water availability and abstraction to allow more water to be taken when more was available. Trading would be made easier, by pre-approving temporary low risk trades. The second option, Water Shares, would be more radical. It would provide abstractors with a share in the available water resource, rather than an absolute amount. In each specified part of a catchment the regulator would assess how much water was required to protect the environment. The remaining "available resource" would then be divided between abstractors as shares with differing reliability.

Furthermore, given that one of the catalysts for reform is that the current system is no longer fit for purpose in the face of future challenges, stakeholders and government were concerned that the new system should be implemented from a sustainable baseline. In other words, that the legacy of damaging or potentially damaging abstraction licences should be addressed before the transition. There is a clear recognition that to fail to do so would result in the sort of unsustainable market seen in the Murray-Darling Basin in Australia. Accordingly, the intention is that unsustainable abstractions will be resolved before implementation of any new system.

For France, the identification of options was strongly influenced by the pursuit of options that would address perceived collective action failures. The main objective of the reform was to anticipate and manage the risk of shortage by apportioning available water among water users. France was looking for an option for allocation reform that would combine the capacity of users to self-regulate and provide an incentive to farmers to act collectively. This option reflected French experience with the development of irrigation and the management of scarcity. In France, experience shows that when water users associations were in place at catchment level, over-allocation was properly managed, as the Chair of the water user association (along with the members) regulates water uses among members in cases of scarcity. Problems emerged in situations where irrigation expansion followed individual plans and decisions, as individual entrepreneurs failed to factor in the consequences of their decisions on the local community.

This experience provided the rationale for France's reliance on OUGCs (*Organismes uniques de gestion collective*) to allocate water and to adjust to shifting circumstances. The objective was to define an institutional framework or policy tool that could provide for the revision of water entitlements, revisit the "first come first served" practice, and grant local farming communities the capacity to self-manage.

In the case of South Africa, the issue of the racial transformation of access to water to address entrenched inequity was on the agenda from the beginning of the process. The policy positions were developed arising from the consultative process and under the strong political guidance of the Minister of Water Affairs. To guide the reform, the Department of Water Affairs, in consultation with stakeholders on a national and regional basis, developed a set of underlying principles. These clarified the overall intention of the process, provided a point of focus for stakeholder consultation processes and established the "policy" or approaches, which would underlie the compulsory licensing process put into place. The elaboration of the principles established allocation reform as a proactive process of actively pursuing the development of water-using enterprises that aligned with national, provincial and local political and development objectives, albeit with a view toward greater efficiency of allocation, rather than a passive process of simply reallocating water.

International experience frequently influences water allocation reform options considered, but leads to different conclusions in different countries

The cases reviewed for this report indicate that international experience is frequently considered in the course of developing policy options. For example, in China, water allocation reform benefitted from keen attention to international experience (tradable water rights in Australia, the EU Water Framework Directive, water quality management in France, etc.). The cases of Israel, England and Wales, and France also illustrate this point. Countries differ in terms of their hydrological endowments, demand positions, institutional arrangements, and preferences for policy approaches, and thus draw different lessons.

In the case of Israel, a review of international experience helped to guide option development. For example, international experience factored into the decision regarding the possibility for private owners of desalination plants to sell water directly to end consumers. An extensive analysis of the advantages and disadvantages was conducted and a comparative study of the situation in OECD countries showed that almost always the local optimum is reached at the cost of the national optimum (Bain Committee, 2010).

To contribute to the options appraisal for the ongoing abstraction licensing reform in England and Wales, the Department for Environment, Food and Rural Affairs (Defra) commissioned reviews of international experience in water allocation. Given that one aim was to make it easier to trade, the Environment Agency also commissioned an international review of other sectors such as fisheries quotas, airport slots and emissions trading. The case studies covered a wide range of experiences of transitions to market-based approaches in sectors either where markets did not previously exist, or where reforms were introduced to improve the way that markets functioned. Parallels with water abstraction were identified, including defining rights that can be traded, managing a gradual transition to markets, and dealing with concerns over market dominance.

France also considered international experience with market instruments for allocation, but drew different conclusions. France has not been impressed by the performance of market instruments for water allocation abroad. In particular, concerns arose about the capacity of tradable water right regimes to factor in environmental and social considerations. Considering issues with water markets abroad on the one hand, and assessing the national needs of France on the other, preference was given to non-tradable water allocation. Thus, France opted for an alternative option, based on an innovative administrative body (the OUGC).

Several factors can constrain the range of policy options considered for water allocation reform

While the range of policy options considered is significantly influenced by the perceived problems with the *status quo*, policy options can be constrained by a number of factors. Limitations imposed by other policy areas (e.g. tax policy) can render some options not legally viable. Strong opposition by stakeholders can also cause certain options to be rejected. In the cases reviewed, certain approaches to pricing provoked particular issues. Conflicts with existing well-entrenched principles for allocation, such as the principle of "first in time, first in line" can constrain the options considered. Broader political or economic considerations provide constraints as well. Finally, institutional path dependency, can raise the cost of improving the flexibility of the allocation regime and reallocating water to higher value uses (see Libecap, 2011). Examples from England and Wales, Alberta, and Israel provide illustrations.

   103

In the course of options development and appraisal in England and Wales, eventually, three preferred options were developed to be ready for formal consultation. One of these – Variable Administered Pricing – was rejected at a relatively early stage. It proposed that the regulator would set a water price according to local water availability. The price would decrease as availability increased, and vice versa in order to protect the environment. During the options appraisal it became clear that this would mean that the charges would be classed as a tax, and so subject to tax policy, which requires that taxes are predictable and stable.

Water pricing provoked strong opposition in Alberta, notably from irrigators, since they perceived it as a type of tax. The consideration of options also generated conflicts with well-entrenched allocation principles. Proportional sharing of water during drought, where all users lose the same proportion of their entitlement, was considered as an option. However, this would be in conflict with the "first in time, first in right" principle, which the government and water users preferred to uphold. Nonetheless, there was recognition among water users that strictly following the priority allocation principle may not be in everyone's best interest in times of drought. An example from the 2001-02 drought illustrated this clearly. In one basin, water entitlements were to be cut off for junior users (with licences newer than 1959). However, this meant that potato growers with senior water entitlements could not send their produce to the processing plant, since it was cut off from water. To address this issue, while retaining the priority allocation principle, the preferred option made it possible for the senior licensees to temporarily assign seniority to some junior licensees.

In Israel, a wide range of different alternatives for reforming allocation and exploring new sources of supply were considered over the years and were studied in many ways. Among the options considered was importing bulk water from Turkey. However, this alternative was discarded after thorough examination due to economic considerations (price of transportation), technological considerations and strategic considerations (creating dependency upon another country).

Economic assessment has proven useful for options appraisal

While it can be difficult to quantify all of the costs and benefits of alternatives, economic assessment has proven to be useful in option appraisal in several countries. For instance, in France, discussion of reform options benefitted from a thorough assessment of the economic and social costs of reduced water licences. Such an assessment was commissioned in the Adour Garonne Basin and Vendée, for example, and generated data which could be used as a reference in the negotiations related to the development of OUGCs.

The case of England and Wales highlights both the challenges of quantifying costs and benefits as well as innovative approaches to deal with them. The impact assessment undertaken in the course of options appraisal highlighted the challenge of quantifying the costs and benefits of the reform options, for several reasons. It requires an understanding of the long-term future scenarios to take into account risks of future water scarcity. It involves the representation of complex trading rules and environmental standards linked to continuously varying water resources. It also involves the representation of short and long-term decision making on water management in the context of uncertainty.

An innovative way of attempting to address these challenges was the use of combined "agent-based" behavioural and hydrological models of four catchments, running in daily steps between 2025 and 2050. Abstractor "agents" were asked to make short and long-term decisions on water management, trading and investment driven by economic and other factors, drawing on behavioural economics. The assessment took account of climate change impacts, and incorporated a range of climate change and socio-economic scenarios. Despite the complexity of the modelling, and its inherent uncertainties, it was valuable in understanding the mechanisms by which policy options might play out, and to present illustrative estimates of likely economic impacts.

The cost-benefit assessment indicated that the reform options provide economic benefits compared with the current system in all scenario combinations. In England, the benefits ranged from about GBP 100 million up to GBP 500 million net present value over 25 years. The impact assessment took account of the way in which particular elements of reform were likely to be implemented estimating the set up costs of GBP 10-16 million for both options, with water shares being more expensive because of its increased complexity. No attempt was made to monetise the benefits to the environment since both options are designed to achieve the same environmental outcomes set in legislation. However, it could have been useful to attempt to do so, providing further insight into the expected net benefits related to the environment of the reforms.

The water allocation reform process

The cases reveal that water allocation reform is not a discrete, time bound process. Instead the process typically extends over many years, even decades, adjusting to changing circumstances. Stakeholder engagement has become common practice for allocation reforms. To mitigate the negative impacts of the reform, negotiating accompanying measures, such as appropriate compensation, and striking compromises among divergent interests are often used to facilitate progress.

Water allocation reforms tend to be iterative processes that extend over many years, even decades

Water allocation reform is not a discrete, time bound process, but instead, the process typically extends over many years, even decades. Reforms tend to be an iterative process in which current "problems" aim to be "fixed" with new measures, which may engender their own problems, which are then addressed by later reforms. The reform process, in many cases, could be seen as a process of maturing the regime, as changes are put into place to address challenges that arise from changing circumstances. Further, while the process of developing of new policies and legislation may be relatively short (2-3 years), an overly technical and complex process to put provisions into practice can stall implementation for decades. A number of cases highlight these dynamics.

Allocation reforms have extended over multiple decades in the Murray-Darling Basin, Australia, in Chile (from the enactment of the Water Code in 1981 to subsequent amendments over the following decades), and in the Yellow River Basin, China, where the allocation regime has been in constant flux over the last 30 years. In South Africa, while the initial expectation was for medium time-frames for the fully implementation of the provisions of the National Water Act adopted in 1998, sixteen years later, there are still significant challenges in implementation (Box 4.5).

Box 4.5. **Factors stalling implementation of water allocation reform in South Africa**

The Water Allocation Reform programme in South Africa recognised early on that getting the pace of reform right was key: move too slowly and you are likely to see radicalisation of policy as the political imperative for redress increases, move too fast and you may threaten the economic value of existing water use, limit the value of improved management of the resource, and increase the likelihood of legal challenges. However, an overly technical and precautionary approach has been taken, and sixteen years since the adoption of the National Water Act (1998), there are still significant challenges in the implementation.

Compulsory Licensing, which had never been implemented anywhere in the world previously, posed particular issues. The concept of Compulsory Licensing was introduced in the Act as a method for the re-allocation of water, primarily from the white minority to the black majority that had been excluded from access to water under Apartheid. This clause enables the minister to call for all water users and potential water users within a specified area to apply for new water use licences, and for the minister then, through a consultative process, to re-allocate the water.

A number of factors have made compulsory licencing difficult to implement. The definition of the reserve for ecological and basic human needs also posed a challenge early on. The Act requires that the ecological and basic human needs reserve be determined prior to the consideration of any licence application. However, there were, initially, no procedures in place for the determination of the ecological reserve. The South African aquatic ecologist community set to work in developing such procedures, facing the challenge of making the transition from a scientific analysis approach to developing assessment tools that would serve the purpose of the Act. The need to determine the ecological reserve for significant water resources in the country prior to the consideration of licence applications significantly delayed the issuing of licences for a number of years. In addition, the translation of the reserve requirements into licence conditions was often difficult. For example, where the reserve determination required a fluctuating flow in the river over different months, where a farmer wanted to construct a simple dam with no mechanisms for releasing such fluctuating flows. In addition, the monitoring of the achievement of the ecological reserve has been weak, and so there is a break in the feedback loop between the issuing of licences and the achievement of the ecological reserve.

The first was that all existing water users were required to register their water use with the Department of Water Affairs (DWA), in order to enable the DWA to have a clear record of who was using how much water and where. However, once a process was introduced to check on the accuracy of this registration and the legality of the water use, it was found that an extremely high proportion of the registered water uses were inaccurate, often irrigation farmers over-registering their water use. This required an intensive process of validating the registration, which is still ongoing. In addition, the failure to put in a requirement that the DWA was informed of any transfer of irrigated land-ownership meant that the registration records were out of date where land had been sold. Since Compulsory Licencing was predicated on having a fairly accurate record of existing water use, this delayed the process.

This rigorous reconciliation process is also intensely legal in nature, which may also underpin the hesitancy the DWA has shown in rolling out the process. Legal challenges could delay the process considerably and the DWA may wish to be very sure of their position before tackling large and difficult compulsory licensing processes.

Source: Schreiner, B. (2014); Quibell, G. (2014).

For Israel, the extensive reform of all of the major pillars of the water allocation system has spanned nearly 20 years. Reforms in Alberta, Canada stemmed from conflicts over low flows as early as the beginning of the 1990s. The allocation reform process in England started in 1997 and is still continuing. For New Mexico, US, after a long build up, allocation reform has been ongoing since 2002 and is still under implementation. Finally, allocation reform has spanned several years in France as well. The 2006 Law on Water and the Aquatic Environment (*Loi sur l'eau et les milieux aquatiques*, LEMA) initiated the reform, which is still ongoing. There have also been numerous delays in implementing the reform, in particular due to farmers asking for further technical and scientific studies, which add to the delays.

Stakeholder engagement has become common practice in water allocation reforms

The case studies attest to the importance of stakeholder engagement[3] in the course of allocation reforms. Stakeholders may be involved at several distinct phases of the process, including the identification and selection of preferred options. While achieving consensus is unlikely, preferred options can sometimes be agreed upon by the majority of stakeholders. In some cases, a thorough review of options can reveal a strong preference for sticking with the *status quo*, despite recognising existing problems. They can also side-track reforms somewhat as preferred solutions can distract from the original aims of the reforms. For instance, in the case of abstraction licensing reforms in England and Wales, certain stakeholders took a view that "markets were the answer, now what is the question?" Stakeholder engagement might not yield the preferred directions towards reform, but can still be valuable for gaining a deeper understanding of the preferences of different water users and spell out what the proposed reform would mean for them. A recent OECD survey on stakeholder engagement indicates that inclusive decision-making leads to better acceptability of decisions on water issues and a greater sense of ownership across the different actors affected (OECD, 2015). Both of these elements are critical for the effective implementation and sustainability of allocation reform.

In Israel, mechanisms encouraging public participation were in place already in the 1959 Law when the Water Council was established. The crisis situation (in the context of the drought) in which the reform took place greatly increased the public media attention to water sector policies. Inquiry committees (with all their collected public testimonies), intensive parliamentary debates on various aspects, and a significant increase in number and influence of NGOs involved in the field are just a few examples of the public's participation in the reform.

In Alberta, Canada, the reform leading to the development of the new Water Act involved public participation at several stages. These included the review of the 1931 Water Resources Act and policies, study visits to neighbouring states undergoing reform, and engagement of technical and legal water specialists to provide expert advice.

Stakeholder engagement also influenced the reforms in New Mexico, United States at several stages. A series of stakeholder forums provided an opportunity to suggest reform options and debate preferences related to these options. The 2002 drought prompted an interim Water and Natural Resources Committee of the New Mexico Legislature to collect stakeholder opinions on how to best reform the water allocation system. Through this process, the preferred option identified (though not a consensus) was to give greater authority to the Office of the State Engineer in terms of administering water rights in locations were court adjudications were still pending (Bossert, 2013).

While this court process was ongoing, a couple of parallel initiatives were taken to investigate the best way to reform the prior appropriation system, which included public hearings. Six stakeholder forums were held in several key cities in which participants were asked to give their views on four suggested reform options. In the end, none of the options was preferred to the *status quo*. The discussion did however help reveal the range of opinions, dilemmas and tensions that exist among the stakeholders (Romero-Wirth and Kelly, 2012). In addition, stakeholders have been engaged in developing voluntary shortage sharing agreements which clarify how water is to be shared during times of drought. These agreements have been largely successful, although mediation has been required in some cases.

Similar to New Mexico, in South Africa, extensive stakeholder engagement was embedded in the reform process, providing multiple opportunities to contribute to and influence the process at several stages. The reform recognised that the successful execution of Compulsory Licensing, while maintaining the rule of law and the right of access to the Courts, was predicated on effective stakeholder participation. Significant objections to the proposed allocation schedule, or significant appeals to the Court could delay the process, and increase its costs significantly, perhaps to the point that it would become moribund. Hence, extensive stakeholder participation was considered essential to the reform process.

In the United Kingdom, it was not unusual for a government consultation on a major reform to be developed behind closed doors and then to be released to a surprised and defensive audience. However, all concerned in the abstraction licensing reform in England and Wales recognised the importance of engaging with those most affected and interested, involving them in the process of development, and using their feedback to refine the options. A series of sector-based and broad-based facilitated workshops helped this process. There was also regular independent expert peer review and analysis, all of which helped with the credibility of the reform.

However, it must be acknowledged that although abstraction reform is intellectually stimulating in itself, for many stakeholders it is peripheral to their core business; what matters to them is how much water they will have, on what terms, and at what level of reliability. Although the "Case for Change" made a compelling scientific argument, which was swiftly accepted, it did not and could not set out what the future might hold for individual abstractors. Further, much of the initial emphasis was on water trading, and this became something of a diversion from the key aim of the benefits of having a reformed, more flexible and dynamic allocation system.

Negotiating accompanying measures and finding compromises to balance divergent interests is essential

To mitigate the negative impacts of the reform, negotiating accompanying measures and striking compromises among divergent interests are often used to facilitate progress. This can include finding a middle ground on contested issues, such as water pricing, or providing compensation in exchange for agreement about reductions to existing water entitlements. Providing an interim period to allow users to adjust to new allocation measures (e.g. prices, changes in entitlements) can also be useful. In some contexts, a powerful central authority can be determinative in resolving disputes. Examples from Israel, France, Australia, South Africa, Alberta, Canada and the Yellow River Basin in China provide illustrations.

An example of finding a middle ground on contested issues can be seen in the case of determining an appropriate level of water pricing for irrigation in Israel. In the course of the allocation reforms, the issue of allocation water for irrigation at very low price was criticised as the State Comptroller determined that "the low selling price of water to the agricultural sector is, to a great extent, the cause of the constant weakening of the water sector" (State Comptroller, 1992). This shift coincided with a rising conflict between the "agro-economic coalition" represented mainly by the Ministry of Agriculture which wished to maintain the subsidised water prices for the agricultural sector, and the "economists coalition" represented mainly by the Ministry of Finance which advocated for an economic-based scale of tariffs as well as a mechanism to regulate the demand and increase the sector's efficiency (Menachem and Gilad, 2013). This dispute caused years of stalemate between these power centers, which left the pricing system for agricultural uses as low and subsidised as possible.

Finally, as part of the efforts to establish the Water Authority, an agreement was reached between the government and the agricultural organisations regarding principles for new water tariffs. The main agreed principle was to relate the sector's freshwater price to the average production and supply costs (including desalination), throughout Israel (Feinerman et al., 2013). In order to assist the agricultural sector due to the expected increase in water tariffs, it was agreed to have an interim period in which incentives by the government would be given to increase water use efficiency.

The reform of the agriculture tariff system came at a timely moment in the lengthy reform process as it coincided with the development of alternative sources of water supply, namely reclaimed water. Encouraging the use of reclaimed water was done by providing wide-scale infrastructure that allowed permanent supply, supported by a financial incentive.

Striking a balance with farmers was also a facilitating factor for France's recent allocation reforms. Over the course of the reform process, farmers would frequently ask for public support to finance the construction and operation of local storage infrastructure to capture abundant winter flows and store water to be used in the dry summer months. Successive governments found a compromise, whereby farmers affected by a reduction of their water licence would be compensated by a licence to build local storage infrastructures. While the compromise was acceptable to farmers, environmental NGOs disputed the measure, claiming that such infrastructures negatively affect landscapes and undermine farmers' incentive to improve water efficiency. As a result, this approach generated significant delays for the reform, as: i) discussions were initiated with several hundred local stakeholders; and ii) economic and ecological consequences of the operation and management of local storages capacities had to be considered.

Accompanying measures were also required to secure sufficient water for the environment in the Murray-Darling Basin, Australia, which entailed significant costs. The government established the Commonwealth Environmental Water Holder to manage the environmental water portfolio and projects to improve the water quality and ecological health in the Murray Darling Basin. AUD 3 billion was committed to buy back water entitlements from irrigators and nearly double this amount (AUD 6 billion) was allocated to upgrade communal and private irrigation infrastructure with the aim to save water for the environment (Cooper et al., 2014).

The thorny issue of compensation has presented challenges for the allocation reform process in South Africa and has been a source of delays in implementation. With regard to compensation, the National Water Act (the Republic of South Africa, 1998) states:

"Any person who has applied for a licence in terms of Section 43 in respect of an existing lawful water use as contemplated in Section 32, and whose application has been refused or who has been granted a licence for a lesser use than the existing lawful water use, resulting in severe prejudice to the economic viability of an undertaking in respect of which the water was beneficially used, may, subject to Subsections (7) and (8), claim compensation for any financial loss suffered in consequence."

The amount of compensation payable is to be determined in accordance with the Constitution, and may not include any reduction in water occasioned by the provision of water for the Reserve, to rectify an over-allocation of water from the resource, or to rectify an unfair or disproportionate water use. The Act gives the Department the right to offer an allocation of water rather than paying the compensation. It also indicates that any claim for compensation has to be lodged with the Water Tribunal set up under the Act which would determine the compensation payable (although the Department of Water Affairs has subsequently disbanded the Water Tribunal, normal recourse to the courts remains).

Although the Agricultural Unions considered challenging the constitutionality of the compensation clauses of the Act, they were advised that a challenge would be more likely to succeed after water had been expropriated without compensation. The lack of legal clarity about the compensation clause led to the reform making all efforts to avoid severe prejudice to the economic viability of enterprises.

While compensation issues are of particular concern in some reforms, in others, a balance was struck between giving greater flexibility within the system and maintaining the security of supply of senior entitlement holders. In the case of Alberta, Canada, measures were put into place to do just that. This measure was the allowing for the possibility to transfer the assignment of priority from one rights holder to another. Basically, this means that a senior user may temporarily assign his seniority to a junior rights holder. Compensation may be paid. Assignments are used mainly to improve the position of junior licensees, whose supply of water is threatened by the priority principle ("first in time, first in line") in times of drought (Adamowicz et al., 2010). At the same time, the security of the allocation of a given entitlement is maintained. Such agreements can be entered into in anticipation of a dry period.

Some municipalities have an existing set of contracts with senior entitlement holders that take effect during a drought, to secure the city's water supply. For example, Calgary, a growing city, fuelled by its energy sector, holds entitlements above its current needs. They have made provisions for water conservation, despite expected population growth. They do not trade their unused water allocation, but share it with other cities during times of scarcity. This mechanism was put into practice by water entitlement holder during the drought of 2002. Some senior users gave their priority over to junior users. This was fully in line with the Water Act of 1999 and not a separate mechanism that users came up with independently. The sharing agreements were facilitated by steering committees made up with community leaders who had interest in communities beyond their own, to ensure that they would negotiate for an agreement that would benefit all.

The case of the Yellow River Basin in China shows that even within a unitary system, preparing an allocation plan posed significant political challenges, requiring resolution of divergent interests. In some contexts, a powerful central authority can be determinative in resolving disputes. Ultimately though, in the absence of a consensus amongst regional governments, the presence of a powerful central government and the capacity of China's State Council were critical in imposing and resolving disputes. So far, adjustments and revisions have been co-ordinated by central administration, usually in response to tensions.

Assessment of water allocation reforms

Given the long timeframes for many water allocation reforms, many cases judged that it is too early to tell if the reform process was successful or not. However, even in the absence of formal and rigorous assessments of reform, evidence of positive developments or disappointment are still available. The cases of Israel, Alberta, Canada, New Mexico, United States and South Africa illustrate positive, mixed or disappointing results.

The impact of water reform in Israel has resulted in significant improvement for the two main drivers that first initiated the reform: addressing significant stress on water resources and improving the economic aspects of water allocation. While the shortage of water remains the key issue in the water sector, and the need to manage the reservoirs between dry and wet seasons and years is the top priority for those responsible for the water sector, authorities have been given enhanced tools to cope with the challenge. Desalination provides a permanent water source, independent from climate variables and a sound basis for domestic consumption. There has also been a noticeable decrease in agriculture's dependency on freshwater, due to extensive use of reclaimed water. Israel today is now a world leader in this area, reusing around 80% of treated wastewater. As a result, in the decade 1997-2007 the use of freshwater in agriculture dropped by 37%, and the use of marginal water grew by 50%. Total water consumption was reduced by 10%, whereas the general efficiency of the agricultural sector with regard to water use increased by 62% (Knesset Research and Information Center, 2008).

The reform has also improved the governance and economic arrangements for water management. The administrative structure of the water sector has changed, with powers concentrated in the hands of the Water Authority, led by the Water Authority Council. The tariff system for all consumer types has improved, since the economic considerations (such as covering the cost for providers) have been given greater importance and the tariff system is simpler and more equal and transparent than it used to be.

In the case of Alberta, Canada, the impact of the allocation reform is generally seen as positive, yet a number of concerns remain. While there has been no formal assessment of the revised regime, there are no major conflicts, court cases or judicial reviews either. Water entitlement holders have keenly embraced the assignment and transfer options introduced by the revision. At latest count, 94 water transfers have been undertaken, most of which are permanent rather than temporary transfers.[4]

However, concern has been expressed about how effective the new rules around trading, transfer and assignment of water rights will be in preventing over-abstraction. The trading scheme intended to include considerations for environmental flows, but this has not been very well-developed and implemented to date. There is a provision in the legislation that stipulates for a given water trade, 10% of the traded volume can be held by the regulator for in-stream uses. This provision contributes to the hesitancy of senior water users to engage in trading. Since there is no effective water metering, it is difficult to determine the 10% and this provision has been inconsistently applied so far.

Furthermore, there is a concern that water markets might activate the unused portions of existing water allocations and thus further strain already overused river systems. In Alberta's *Water for Life* strategy, the government expressed its commitment to evaluate the merits of using economic instruments, such as water pricing, to meet key water conservation objectives (de Loë et al., 2007). A study based on a network model of the entire river basin in southern Alberta found that the relative efficiency gains from introducing market pricing could range from under 3% in a year of surplus flows to more than 15% in a drought year (Mahan et al., 2002).

In a similar vein, the informal assessment of the reform in New Mexico highlights current perceptions of the benefits and drawbacks of Active Water Resources Management[5] (AWRM) put into place to facilitate re-allocation of water. On the positive side, conflicts can be resolved more quickly compared to the lengthy court adjudication procedure. The use of voluntary and temporary transactions through water leases or shortage sharing agreements has also been facilitated. A good example of this is from southeast New Mexico where there is currently an oil and gas boom and farmers lease their water rights to oil and gas companies.

Another development is the facilitation of in-stream flows protection. Unless denied by the State Engineer or challenged in court, any individual (including the State) can lease or purchase existing consumptive water rights and convert them for environmental use while maintaining ownership of the rights (Scarborough, 2010). An additional feature of the new water allocation regime is that historical water users have been given a say in water transfers. *Acequias*[6] or qualifying ditch companies are allowed to adopt bylaws requiring their approval as a condition to surface water transfers (Western Governors' Association, 2012).

On the negative side, AWRM is a resource dependent procedure. It requires more human resources since water masters in the field are needed in each basin. It is also dependent on metering of water use and many users do not currently have meters and are not very keen on installing them. The Office of the State Engineer can require meter installation by law, and has funds to do so, but prefers users to do it voluntarily, as a legal process would be required to compel users to do it. Further, some have negative views of the prerogative of OSE to not be obliged to strictly follow the principle of "first in time, first in right" but instead put more focus on interpreting the principle of "beneficial use" (a principle which is indicated but not explicitly defined in the Water Code). This gives more weight to uses that produce higher economic values, such as industrial and urban uses.

In contrast to positive or mixed assessments of water reform in the cases above, for South Africa, the results overall have been disappointing. There have been a number of independent assessments the water allocation reform process.[7] For example, Muller (2013) argues that it is too early to draw lessons about the success of the water reform in South Africa, partly because hydrological favourable conditions (i.e. the lack of a major, nationwide drought) since 1994 have not 'tested' the system. Secondly, he argues that demand for water from the agricultural sector has been relatively stagnant due to slow progress in rural land reform.

However, other observations are more critical of the process. Merrey (2008) argues that the focus on large scale users militated against enabling poor users to get access to water for productive purposes. He also refers to the lack of an integrated approach across departments, particularly in relation to the land, agricultural and mining sectors. Merrey is also critical of "the optimism about using water as a lever to achieve social and economic reforms" (2008). In his view, this was unrealistic due to the cautious, technocratic approach taken to implementation and even more importantly, the lack of a alignment and integration with land and agricultural reforms. Movik (2009) argues that what was initially an extremely political approach to water allocation reform ended up in a technicist approach with implementation that was too complex in a context of limited human resources for the purpose resulting in stalled delivery.

What is clear is that the water allocation reform programme in South Africa has failed to meet expectations, with very limited water allocations having been made to black South Africans after 20 years of democracy, and little evidence of effective implementation and monitoring of the ecological reserve.

Conclusion

The case studies illustrated in this chapter reveal insights and lesson learned on the reform of allocation regimes. The experience of allocation reform is unique to its time and place, but nevertheless, common themes emerge across the case studies relating to reform drivers, key elements of the reform process and implementation.

Allocation reform is often driven by an accumulation of factors, with concerns about water scarcity and insufficient water for ecosystems as commonly cited drivers. These concerns are often combined with broader political or structural reforms to improve the economic efficiency and equity of allocation arrangements. Several cases illustrate how droughts can provide a salient, visible event to trigger action.

Water allocation reform is essentially a socio-economic and political process, and adopting an overly technical approach can result in delays to the water reform programme. It also benefits those who have the greatest capacity for engaging with strongly technical and legal processes, which tend to be the already privileged. The reform process allows for ample opportunities for participation and negotiation, which can make it difficult to maintain control over a tight implementation plan and schedule. However, the negotiations can be an effective means to devise compromises and appropriate compensation measures to mitigate the potential negative impacts of the reform given that the reform process is technically, politically and legally challenging. This must be recognised from the outset, and as such extensive stakeholder engagement is a must. The case studies illustrate the various forms such participation can take at various stages throughout the process.

In terms of developing options for reform, the case studies illustrate how reform options are significantly shaped by the perceived shortcomings of the existing system. Hence, the way in which the "problem" of allocation is defined strongly influences the potential solutions considered. International experience is frequently drawn on in the process of formulating reform options, but different countries draw different lessons, in particular with regard to the desirability of market-based approaches. The feasibility and desirability of options can be constrained by various factors, such as legal constraints and strong stakeholder opposition. The cases of France and England and Wales illustrate how economic assessment has proved useful to aid the selection of options, even if it may be difficult to quantify all the costs and benefits of alternatives.

The cases also highlight lessons relating to the sequencing of reform and approaches that can help to minimise unintended consequences. Allocation reforms require a holistic view of the system, since the reform of only one aspect of allocation can have unintended impacts on other parts of the system. In terms of sequencing, the cases attest to the importance of determining a sustainable baseline (to determine how much water is available for allocation), before introducing changes to an allocation regime. Similarly, ensuring the hydrological integrity in the allocation system by accounting for return flows is important, especially when introducing trading that can shift water among uses with different rates of consumption. For instance, if entitlements are based on total withdrawals and a transfer is made from a user that has low consumptive use to one that has high consumptive use, return flows to the basin are reduced, putting pressure on the sustainability of the resource. The potential impact of reforms on non-exercised water entitlements should be fully considered in advance as well.

Multiple cases attest to the desirability of increasing the flexibility that water users have to adjust to changing conditions and reallocation water use among themselves. Several cases have used water trading and markets to achieve this. The case of France illustrates an institutional arrangement that allows for collective bargaining to try to achieve a similar effect. However, institutional path dependency can raise the costs of improving the flexibility of allocation arrangements to respond to uncertainty and reallocating water to higher value uses. Allocation arrangements, such as prior appropriation in the United States, introduced to meet historical objectives, constrain contemporary economic opportunities and cannot be easily modified or replaced *ex post* (Liebert, 2011). Finally, the recognition that policy and legislative reform is only as good as the ability to implement it, calls for appropriate support, and capacity for implementation.

Notes

1. For example, according to Riestra (2008) among the 15 000 m^3/s granted non-consumptive uses, right only 2 800 m^3/s was actually being exercised.

2. For example, the non-use tariff for the shortest time period (0-5 years) is determined by a pre-set rule. This charge doubles if the right is not used during 6-10 years and quadrupled from 10 years onward (Madden, 2010).

3. In this context, stakeholder engagement is defined as the process by which any person or group who has an interest or stake in a water-related topic may be directly or indirectly affected by water policy and/or has the ability to influence the outcome (positively or negatively) is involved in the decision-making processes (OECD, 2015).

4. Some transfers simply changed the piece of land to which the water entitlement was attached, under the same owner. A limited number of trades have occurred in the South Saskatchewan basin. Trades between different types of users (e.g. agricultural to domestic) have been rare.

5. In essence, Active Water Resource Management strengthens the authority of the State Engineer in New Mexico to temporarily reallocate water by establishing "water master districts", where a water master is appointed in charge of administering water rights. Rights that are junior to an "administration date" set by the State Engineer can be curtailed. Uses that are not prioritised can apply for a "replacement plan", for a maximum duration of two years. Water users are also allowed to submit joint "replacement plans" or to find voluntary shortage sharing agreements, through negotiations (Romero-Wirth and Kelly, 2012).

6. Associations managing communal irrigation canals.

7. See Merry (2008), Movik (2009), and Muller (2013). In addition, in 2013 the minister requested a policy review in tandem with a review of the national water legislation.

References

Adamowicz, W.L., D. Percy and M. Weber (2010), *Alberta's Water Resource Allocation and Management System: A Review of the Current Water Resource Allocation System in Alberta*, University of Alberta/ Alberta Innovates, Edmonton, Alberta.

Alberta Environment (2003), *Water for Life Strategy*, Ministry of Environment and Sustainable Resource Development, Edmonton, Alberta.

Arlozorov Committee (1997), Report of the Committee for Examining the Management of Water Supply in Israel, Ministry of Infrastructures, Jerusalem.

Bain Committee (2010), National Committee report on the management of the water sector in Israel, a special report submitted to the Israeli Parliament (Knesset), Jerusalem, *http://elyon1.court.gov.il/ heb/mayim/Hodaot/00407510.pdf* (accessed 15 June 2014).

Barker, I. (2014), "Water Resources Allocation: Reform in England and Wales", *Water Policy International Ltd*, background paper for the OECD project on water resources allocation, (unpublished).

Bauer, C. (2004), "Siren Song: Chilean Water Law as a Model for International Reform", Resources for the Future Press, Washington, DC.

Bitran, E. and R. Saez (1994), "Privatization and Regulation in Chile", in Bosworth, B.P., R. Dornbusch and R. Laban (eds.), *The Chilean Economy: Policy Lessons and Challenges*, The Brookings Institution, Washington, DC.

Bossert, P. (2013), "Active Water Resource Management", in *Water Matters! – Background on Selected, Water Issues for Members of the 51st New Mexico State Legislature 1st Session* (Chapter 9), New Mexico University School of Law, Albuquerque, *http://uttoncenter.unm.edu/publications.php* (accessed 5 May 2014).

Büchi, H. (1993), *La transformacion economica de Chile. Del estatismo a la libertad económica*, Bogota, Norma.

Cooper, B., L. Crase and N. Pawsey (2014), *Best Practice Pricing Principles and the Politics of Water Pricing, Agricultural Water Management*, Vol. 145, Elsevier, B.V., pp. 92-97.

de Loë, R.C. et al. (2007), *Water Allocation and Water Security in Canada: Initiating a Policy Dialogue for the 21st Century*, Report prepared for the Walter and Duncan Gordon Foundation, Guelph, ON: Guelph Water Management Group, University of Guelph, Ontario.

Donoso Harris, G. (2011), "The Chilean Water Allocation Mechanism, established in its Water Code of 1981", European Commission, *www.feem-project.net/epiwater/docs/d32-d6-1/CS30_Chile.pdf* (accessed 10 January 2014).

Donoso Harris, G. (2003), *Water Markets: Case Study of Chile's 1981 Water Code, Global Water Partnership, www.eclac.cl/drni/proyectos/samtac/inch01603.pdf* (accessed 18 November 2013).

Feinerman, E., H. Frenkel and U. Shani (2013), "The Water Authority: The Impetus for its Establishment, Its Objectives, Accomplishments and the Challenges Facing It", in Feitelson, N.E., *The Four Eras of Israeli Water Policies*, in N. Becker (ed.), *Water Policy in Israel: Context, Issues and Options*, Springer Science + Business Media, Dordrecht.

Feitelson, N.E., *The Four Eras of Israeli Water Policies*, in N. Becker (ed.), *Water Policy in Israel: Context, Issues and Options*, Springer Science + Business Media, Dordrecht.

Government of Israel (2013), "Preface", *Budget and Economic Policy for 2013-14, www.mof.gov.il/BudgetSite/ StateBudget/Budget2013_2014/Lists/List2/Attachments/4/MediniyutClacalit.pdf* (accessed 15 June 2014).

Hearne, R. and G. Donoso Harris (2005), "Water Institutional Reforms in Chile", *Water Policy*, Vol. 7(1), IWA, pp. 53-69.

Knesset Center for Research and Information report (2008), "The crises in Israel's Water Sector, the implementation of the parliamentary inquiry committee for the water sector and sustainable development" (Hebrew), submitted to the Knesset State Control Committee, 27 July 2008, *www.knesset.gov.il/mmm/data/pdf/m02118.pdf*.

Libecap, G. (2011), *Institutional Path Dependence in Climate Adaptation: Coman's "Some Unsettled Problems of Irrigation"*, Vol. 101, *American Economic Review*, February 2011, pp. 64-80, *http://dx.doi.org/10.1257/ aer.101.1.64*.

Madden, E. (2010), "Chilean Water Policy: Transaction Costs and the Importance of Geography", *Paper*, No. 65, Dietrich College Honors These, *http://repository.cmu.edu/hsshonors/65* (accessed November 2013).

Magen Committee (2002), Parliamentary Commission on the water economy headed by M.K. David Magen, *www.knesset.gov.iul/committees/heb/docs/vaadat_chakira_mayim.html*.

Mahan, R.C., T.M. Horbulyk and J.G. Rowse (2002), "Market mechanisms and the efficient allocation of surface water resources in southern Alberta", *Socio-Economic Planning Sciences*, No. 36, pp. 25-49.

Menachem, G. and S. Gilad (2013), "Israel's Water Policy 1980-2000: Advocacy Coalitions, Policy Stalemate, and Policy Change", in N. Becker (ed.), *Water Policy in Israel: Context, Issues and Options*, Springer Science + Business Media, Dordrecht.

Merrey, D. (2008), "When Good Intentions are not Enough: Is Water Reform a Realistic Entry Point to Achieve Social and Economic Equity?", *www.waternetonline.org/Symposium/9/full%20papers/ws/Merrer,%20DJ.doc* (accessed 20 August 2014).

Movik, S. (2009), "The Dynamics and Discourses of Water Allocation Reform in South Africa", *http://opendocs.ids.ac.uk/opendocs/handle/123456789/2287#.U-iZuvmSyQc* (accessed 20 August 2014).

Muller, M. (2013), "Lesson from South Africa on the management and development of water resources for inclusive and sustainable growth", *www.academia.edu/1564589/Lessons_from_South_Africa_on_the_management_and_development_of_water_resources_for_inclusive_and_sustainable_growth* (accessed 20 August 2014).

OECD (2015), *Stakeholder Engagement for Inclusive Water Governance*, OECD Studies on Water, OECD Publishing, Paris, *http://dx.doi.org/10.1787/9789264231122-en*.

OECD (2010), *Sustainable Management of Water Resources in Agriculture*, OECD Studies on Water, OECD Publishing, Paris, *http://dx.doi.org/10.1787/9789264083578-en*.

Pena, H. (2010), "Copiapo; realidades, desafio y lecciones. Presentación para 'existe sobreexploatacion del agua en chile?'", seminario nacional AHLSUD Capitulo Chileno A.G.

Pena, H., M. Luraschi and S. Valenzuela (2004), "Water, Development, and Public Policies – Strategies for the inclusion of Water in Sustainable Development", South American Technical Advisory Committee, Global Water Partnership, Santiago de Chile, Chile.

Quibell, G. (2014), "South Africa's water allocation reform process and its application to the Inkomati transboundary basin", background paper for the OECD project on water resources allocation (unpublished).

Republic of South Africa (1998), *National Water Act*, Act No. 36 of 1998, *www.dwaf.gov.za/Documents/Legislature/nw_act/NWA.pdf*.

Riestra, F. (2008), "Pago de patente por no uso de derechos de aprovechamiento", Direccion General de Aguas (DGA), Santiago de Chile, Chile.

Romero-Wirth, C. and S. Kelly (2012), *Water Rights Management In New Mexico and Along the Middle Rio Grande: Is AWRM Sufficient?*, *http://uttoncenter.unm.edu/publications.php* (accessed 5 May 2014).

Scarborough, B. (2010), "Environmental Water Markets: Restoring streams through trade", *PERC Policy Series*, No. 46, Bozeman, Montana.

Schreiner, B. (2014), "Water resources allocation reform in South Africa: Case study", background paper for the OECD project on water resources allocation (unpublished), Pegasys Consulting.

State Comptroller (1992), "Managing the Water Sector in Israel", *Report* 41A, State Comptroller, Jerusalem.

State Comptroller (1990), "Report on the Water Sector Management in Israel", State Comptroller, Jerusalem, *http://old.mevaker.gov.il/serve/contentTree.asp?bookid=345* (accessed 15 June 2014).

Western Governors' Association (2012), "Water Transfers in the West: Projects, Trends and Leading Practices in Voluntary Water Trading", The Western States Governors' Association, Western States Water Council, Denver, Colorado, *www.westernstateswater.org/wp-content/uploads/2012/12/Water_Transfers_in_the_West_2012.pdf* (accessed 20 August 2014).

Chapter 5

A "Health Check" for Water Resources Allocation

This chapter sets out a "health check" for improving the performance of allocation regimes. It builds on both the analytical framework for allocation developed in Chapter 2 as well as the practical experience of a range of examples collected via the OECD Survey of Water Resources Allocation analysed in Chapter 3. It uses a series of "checks" to identify whether key elements of an allocation regime are in place and how their performance could be improved. In some cases, several options for the design of elements are proposed.

Key messages

- The overall policy guidance for improving water allocation arrangements is presented in this chapter as a ***"Health Check" for Water Resources Allocation***. It is a tool to review current allocation arrangements to check whether the elements of a well-designed allocation regime are in place and to identify areas for potential improvement.

- Since the risk of shortage is dynamic, in both the short-run and the long-run, **a well-designed allocation** should have two key characteristics: it should be **robust** by performing well under both typical and extreme conditions and demonstrate **adaptive efficiency** with the capacity to adjust to changing conditions at least cost over time.

- Allocation regimes need to be tailored to specific conditions. In the early stages of developing a water resource, or where the risk of shortage is low, a relatively simple design can be used with decisions made conservatively to avoid over-allocation and over-use. As scarcity increases and the value of water use rises, the **benefits of a more elaborate allocation regime increases**.

Many current water allocation regimes have evolved in a piecemeal fashion over time and are strongly conditioned by historical water usage patterns. As a result, they are usually not well-equipped to deal with mounting pressures on the resource, such as changes in water demand and climate change, or shifts in societal pressures, such as increasing value placed on environmental flows to support ecosystem services. The challenges for allocation are aggravated by the entrenchment of weak water policies (under-pricing water or lack of regulating use), which contribute to structural water scarcity, increasing the risk of shortage for users and for the environment. The significant path dependency of allocation regimes, which manifests in laws and policies, and even in the design and operational rules of existing water infrastructures, can make adjustments to allocation arrangements very contentious and costly. The result is that many allocation regimes are no longer "fit for purpose" and require adjustments in order to adapt to changing conditions.

To address this challenge, policy guidance for improving allocation arrangements is presented in this chapter as a *"Health Check" for Water Resources Allocation*.[1] The Health Check is designed as a tool to review current allocation arrangements in a specific context to check whether the elements of a well-designed allocation regime are in place and to identify areas for potential improvement. Since the risk of shortage is dynamic, in both the short-run and the long-run, a well-designed allocation should have two key characteristics: it should be robust by performing well under both typical and extreme conditions and demonstrate adaptive efficiency[2] with the capacity to adjust to changing conditions at least cost over time.

The Health Check can be applied to various scales of water governance, depending on the context. For example, it can be used at the national, provincial/state, or river basin level, or used for a specific irrigation district. Used iteratively, the Health Check can be used to drive further improvements and refinements to more fully reap the benefits of a well-designed allocation regime and to adjust to changing circumstances. Box 5.1 provides a summary of the Health Check. Each of the elements are discussed in detail in this chapter.

An allocation regime should be tailored to specific conditions. In general, as the risk of shortage increases, the benefits of a more elaborate allocation regime increases. In the early stages of developing a water resource, or when the risk of shortage is low (e.g. due to water abundance), a relatively simple allocation regime can be used with decisions made conservatively to avoid over-allocation and over-use. However, the basic building blocks of a robust regime should still be put into place at an early stage, which can allow for adjustment at least cost as needed over time. As scarcity increases and the value of water use rises, the case for the introduction of a more elaborate allocation regime based on clear, adjustable limits on abstraction and clear, legally defined volumetric entitlements increases. When water over-allocation and/or over-use already exists, there is an opportunity to use the characteristics of a more elaborate allocation regime to reduce the extent of the problem and bring use in line with sustainable limits.

Scarcity conditions can be understood as a continuum. Water scarcity is typically a slow onset risk resulting from a water deficit that accumulates over time, in contrast to a sudden impact caused by a singular event, such as a storm or flood (OECD, 2013a). No or low scarcity conditions exist where there is general water abundance and a low incidence of drought. As long as there is sufficient water available to meet both latent and current demand sustainably,[3] water resources are neither over-allocated nor over-used. Moderate scarcity conditions exist where scarcity is either an emerging threat (e.g. due to increased

---

Box 5.1. **A "Health Check" for Water Resources Allocation**

Check 1. Are there accountability mechanisms in place for the management of water allocation that are effective at a catchment or basin scale?

Check 2. Is there a clear legal status for all water resources (surface and ground water and alternative sources of supply)?

Check 3. Is the availability of water resources (surface water, groundwater and alternative sources of supply) and possible scarcity well-understood?

Check 4. Is there an abstraction limit ("cap") that reflects *in situ* requirements and sustainable use?

Check 5. Is there an effective approach to enable efficient and fair management of the risk of shortage that ensures water for essential uses?

Check 6. Are adequate arrangements in place for dealing with exceptional circumstances (such as drought or severe pollution events)?

Check 7. Is there a process for dealing with new entrants and for increasing or varying existing entitlements?

Check 8. Are there effective mechanisms for monitoring and enforcement, with clear and legally robust sanctions?

Check 9. Are water infrastructures in place to store, treat and deliver water in order for the allocation regime to function effectively?

Check 10. Is there policy coherence across sectors that affect water resources allocation?

Check 11. Is there a clear legal definition of water entitlements?

Check 12. Are appropriate abstraction charges in place for all users that reflect the impact of the abstraction on resource availability for other users and the environment?

Check 13. Are obligations related to return flows and discharges properly specified and enforced?

Check 14. Does the system allow water users to reallocate water among themselves to improve the allocative efficiency of the regime?

---

demand, increased variability or incidence of drought) or a periodic and localised threat to water supply. In this case, water resources may be over-allocated, but not yet over-used. Finally, severe scarcity conditions exist where scarcity is a chronic issue due to significant climatic variability, incidence of drought, or situations where demand outstrips available supply, thus generating scarcity conditions even in average years (e.g. structural scarcity). In these cases, the water resource is often either over-allocated or over-used, or both. Figure 5.1 provides an illustration of the scarcity spectrum.

Figure 5.1. **Water scarcity spectrum**

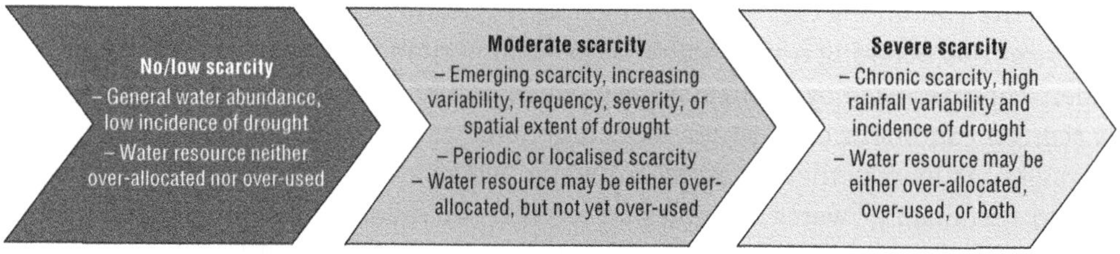

Over time, countries and regions may move along this spectrum. An increasingly elaborate allocation regime is appropriate as scarcity conditions become more acute. Many of the illustrations in boxes in this chapter would be most suitable for moderate or severe scarcity conditions. However, the presence of no or low scarcity conditions does not mean that attention should not be paid to the design of an allocation regime. The building blocks of a well-designed regime should still be put into place and designed in a way that would encourage adaptive efficiency should conditions evolve. This includes avoiding design elements with a high level of path dependency, which are costly to change (see discussion in Chapter 4).

As discussed throughout this report, allocation regimes consist of a combination of policies, mechanisms, and governance arrangements (entitlements, licenses, permits, etc.) that are used to determine who is allowed to abstract water from a resource pool, how much may be taken and when, how much water must be returned (and in what quality), and the conditions associated with the use of this water. Some elements are best managed at the system level and others at the user level. The robustness and adaptive efficiency of an allocation regime can be improved by unbundling the various elements and using separate instruments to pursue various objectives (see Box 2.4). The system can also introduce measures to allow water users to reallocate water among themselves to improve the allocative efficiency of the regime. For example, market instruments such as water entitlement exchange centres or entitlement transfer contracts can help alleviate extreme scarcity conditions, as was the case in Spain during the drought period 2005-08.

System level elements of a water allocation regime

This section focuses on options for the management of system-wide challenges in regions where water scarcity either is or is expected to be a constraint. System level elements are those that are most efficiently and equitably dealt with at the scale of the water resource, whether it is the basin, catchment, river, stream or aquifer. Usually, they take the form of conditions expressed in water sharing plans and other similar documents that determine how system wide decisions are taken. Those that apply to all water resources can be contained in regional and national legislation.

Check 1. Are there accountability mechanisms in place for the management of water allocation that are effective at a catchment or basin scale?

- Authorities and organisations responsible for allocation should have well-defined roles and accountability mechanisms that actually work in practice, as well as sufficient resources (financial and otherwise) to execute their function. A **River Basin Management Plan** (or other similar planning instrument) that has the status of a statutory instrument that must be followed can be used to set out a clear framework for allocation. A clear and **transparent process** should be in place to facilitate stakeholder engagement in the definition of the sequence of priority uses and other key allocation decisions.

Check 2. Is there a clear legal status for all water resources (surface and ground water, as well as alternative sources of supply)?

- A **clear legal status** should be in place for all types of water resources (surface and ground water, as well as alternative sources of supply, such as treated wastewater). This status needs to define whether the resources are publicly or privately owned, or in cases where there is no ownership of water resources, *per se*, who has the authority to determine access to the resource. In cases where a plurality of legal regimes relating to water resources exists, **competing claims need to be clarified**, to avoid "allocation by litigation" or "allocation by adjudication". Any contradictory and overlapping legal arrangements relating to the ownership of the resource itself as well as legal entitlement to access and use water resources should be clarified. A clear legal definition of waste water as a resource would be useful to promote waste water re-use.

- Obligations under **international agreements** related to **transboundary water resources** should also be clearly specified. (See Box 2.3 on the Albufeira Convention,[4] a good example of clearly defined transboundary co-operation between Spain and Portugal.)

Check 3. Is the availability of water resources (surface and ground water, as well as alternative sources of supply) and possible scarcity well-understood?

- A **robust scientific basis** is needed to identify the available water resources (surface and ground water, as well as alternative sources of supply, such as treated wastewater), understand how they may be interconnected, when they are available (seasonality) and how they may change over time. This requires an assessment of water resources at the relevant scale for allocation with a view to determining which water resources are currently under pressure from water scarcity or are likely to be so in the future. This may be particularly challenging for groundwater resources, for which there seems to be a relatively low or insufficient level of knowledge (see OECD, 2015). The **comprehensiveness of the assessment** should correspond to the degree of scarcity conditions, with water resources under greater pressure deserving more in-depth assessment as compared to resources where scarcity is not yet an issue and not expected to be in the near future. In general, it is not possible, nor is it necessary, to obtain complete knowledge of water systems. Instead, the aim should be to acquire sufficient knowledge of the available water resources in order to make appropriate and tailored decisions (see Box 2.5 for examples from Spain and France). The information should be made publically available.

- Managing **system interconnectivity** is essential for ensuring the hydrological integrity of the system. For instance, careful consideration needs to be given to the impact of groundwater bores located next to a river. In such situations, extraction from the bore may in fact be extraction from a river. In order to avoid double-counting that will result in over-use in such circumstances, the amount of water that may be taken from the river needs to decrease and the amount of water taken from the aquifer can increase. Arrangements need to be in place to adjust for changes in flows between groundwater and surface water systems. Well-designed regimes can also encourage water users to seek out opportunities to take surface water, for example, and store it in an aquifer.

- Where economically viable, **non-conventional water sources**, such as treated wastewater and desalinated water, could be considered as an alternative to conventional freshwater supplies during drought periods or as a complement to conventional supplies during normal periods.

Check 4. Is there an abstraction limit ("cap") that reflects in situ requirements and sustainable use?

- Defining a **limit on the maximum volume or proportion** of water that can be abstracted from a system is arguably the most difficult and yet most important challenge in managing water scarcity. Setting an abstraction limit requires consideration of the amount of water that must be put aside to meet *in situ* requirements and downstream obligations (in the case of surface water), as well as needs for non-consumptive uses and environmental needs. Both policy-related and technical limitations on the quantity of water available for sustainable use need to be recognised. Technical restrictions limit economically viable water use. Policy-related limitations may include obligations related to minimum flows for transboundary rivers, environmental flows, or indigenous rights. Environmental flows are needed to meet ecological objectives, but also to protect downstream users. The definition of the "cap" also needs to consider quality (including temperature) requirements. Different options can be considered to define how these flows are treated within an allocation regime (see Box 2.6).

- Two types of abstraction limits are needed:

  - A **long-term limit** that defines the maximum volume of water that can be abstracted at any point in time. Once this limit has been fully allocated, no new entitlements should be issued unless the process is accompanied by an arrangement that reduces someone else's entitlement by an equivalent amount. A **mechanism to adjust the long-term limit** is needed for adaptive management. This is especially the case in regions with high rainfall variability and expectations of adverse impacts of climate change, but can also relate to the need adapt to other drivers of change, for instance as a result of new scientific evidence about ecosystem needs. The long-term limit can be used to guide strategic water-dependent investments.

  - A **short-term limit** on the amount of water that can be taken at a particular point in time. In addition to limits on the maximum amount of water that can be taken over the long-term, in most regions, it is also necessary to be able to **adjust the amount of water that can be taken** within a season. In the definition of a short-term limit, one must take into account the time needed for water users (including irrigators) to adjust. For instance, the short-term limit needs to be determined and announced in advance of the planting season if it is to influence cropping decisions.

Check 5. Is there an effective approach to enable efficient and fair management of the risk of shortage that ensures water for essential uses?

- **"Essential"** water uses should be defined and assured the highest priority. This includes drinking water for humans and national security uses (such as flood protection or cooling for nuclear power plants). Water needs for the environment that reflect the dynamic flow regime should also be secured.

- Efficiency in allocation can be improved if a suite of entitlement classes is established and users are allowed to invest in a **portfolio of reliability classes**. This allows users to manage the degree of supply risk they wish to take. In order to allow for the efficient management of long-term supply risks, at least two priority classes are needed: a high priority and a low priority class. Most allocation regimes already have three or four priority classes implicit in their current structure, so it may be easier to establish more classes than fewer. With such an approach, allocations are first made to the high priority class. Once these entitlement holders have received 100% of their entitlement, allocations are then made in turn to each of the lower classes according to their priority. Provided that there are at least two priority classes and shares in them are freely tradable, investment risks can be managed by varying the proportion of each share type held.

- As an alternative, "second best" option, a **sequence of priority uses** can be established that is applied in periods of scarcity, or used to guide the allocation of entitlements in "normal" periods. Priority regime banning is a pragmatic approach and works as a short-term strategy in cases where shortage incidents occur infrequently. It has a relatively low cost of implementation and there is little need for investment in the development of elaborate administrative arrangements. However, it can be a source of "lock-in" and make managing tensions among various users more difficult. It also places the risk of shortage disproportionately on users designated as "low priority". If a sequence of priority uses is in place, it should generally reflect the ranking of relative value of water across various uses. The determination of a sequence of priority uses needs to take into account the sustainable use of the resource.

Check 6. Are there adequate arrangements in place for dealing with exceptional circumstances (such as drought or severe pollution events)?

- The conditions that constitute an "exceptional circumstance",[5] such as a drought or severe pollution event, need to be clearly specified. Stakeholders should be involved in the process of determining what constitutes exceptional circumstances. A **responsible authority** that has authority to declare an exceptional circumstance and manage the response needs to be designated. Water users need to be informed regularly about the developments relating to exceptional circumstances and the how they will be affected by the response. The more advance warning that users can be provided, the more opportunity that they will have to adjust their behaviour and effectively manage their risk of shortage. Advance warning can encourage the conservation of resources and help to minimise the impact of their abstraction on the environment.

Check 7. Is there a process for dealing with new entrants and for increasing or varying existing entitlements?

- In cases where the defined resource pool is fully allocated, the resource should be considered "closed". Once access to the resource is closed, the only way a new entrant may secure an interest in abstracting water from the resource or an existing use may expand an existing entitlement is to ensure that **another user foregoes use of an equivalent amount**, thereby transferring the water entitlement to the new entrant or the existing user expanding an entitlement.

Check 8. Are there effective mechanisms for monitoring and enforcement, with clear and legally robust sanctions?

- An appropriate level of **monitoring** of the resource, ecosystem requirements, abstractions, and discharges that reflects the level of pressure on the water resource needs to be in place. Losses in the water distribution network should be monitored as well, to inform decisions about leakage reduction. Rigorous monitoring requires an assessment of the volume of water being taken by each user. This requires the installation of meters, meter reading, and accounting protocols. Appropriate accounting arrangements that track water use and consumption, as well as leases and trades where permitted, need to be in place to support the monitoring of water use and water entitlements.

- Appropriate **sanctions**, such as fines or curtailment of water entitlements need to be in place and applied as required.

- **Uncontrolled uses** and any significant interception need to be periodically reviewed to gauge their potential impact on the integrity of the system. When uncontrolled uses and significant interceptions begin to have a significant impact on the water system, they must be brought into the formal water entitlement system. This sends a clear signal to existing entitlement holders that the expansion of these uses will not undermine the efficiency of any investments they have made.

Check 9. Are water infrastructures in place to store, treat and deliver water in order for the allocation regime to function effectively?

- Adequate water **infrastructures** are needed to store, treat and deliver water to various uses and users and a lack thereof can place constraints on the flexibility of allocation regimes. Pervasive uncertainty about future climatic conditions increases the value of scalable and "adaptable" water infrastructure, which are typically capital intensive and long-lived, often with high sensitivity to climate. Scalable, modular, less capital intensive approaches to infrastructures (including grey or "green") can provide an additional "option" value, as they are more easily adjusted to future conditions (OECD, 2013b).

- Water infrastructures and their **operational rules** need to be managed in a way that effectively contributes to the aims of the allocation regime. When appropriately managed, multi-purpose infrastructures can contribute to allocative efficiency.

Check 10. Is there policy coherence across sectors that affect water resources allocation?

- The existing policy settings related to water resources management as well as water-related sectors, such as agriculture and energy, need to be coherent. Even a well-designed allocation regime can be undermined by perverse incentives in other sectors, such as subsidies that encourage over-consumption of water resources or pollution that degrades water quality.

User level elements of a water allocation regime

User-scale elements refer to those factors that are most efficiently and equitably dealt with by specifying the arrangements that apply to an individual user. Typically these take the form of arrangements specified in water entitlements (also referred to as permits, abstraction licenses, etc.).

Check 11. Is there a clear legal definition of water entitlements?

- Well-functioning allocation regimes need to have **clear, quantified, legally defined water entitlements**. Options for defining how users can access water and how much they are allowed to take range from a requirement that the user own land adjacent to a river to a requirement that all abstractions be metered and that the amount abstracted always be inferior to that credited to a water account. The choice of the most appropriate way to define these entitlements depends on the value that access to water brings to an economy and the contribution it makes to the environment. To improve the flexibility of the allocation regime, water entitlements can be **unbundled** from land titles. Flexibility at the system level can be balanced with security at the user level by managing water entitlements separately from allocations in a particular season.

- There are benefits to defining water entitlements as a **proportion**, or shares, of the available resource pool (as opposed to an absolute volume). This approach allows for flexibility to respond to changing conditions without having to pay compensation for adjusting water entitlements. This approach is also consistent with the full assignment of risk. Conversion from a volumetric or seniority regime to a proportional regime is possible, although it may be challenging.

- Water entitlements must be defined for an **appropriate duration**, with a clear, reasonable expectation for renewal. This could be a fixed period of time, or water entitlements could be defined in perpetuity. The longer the entitlement is granted for, the more it will encourage long-term investment in water-related activities. Uses that require significant investment to benefit from the water entitlement merit a longer duration.

Check 12. Are appropriate abstraction charges in place for all users that reflect the impact of the abstraction on resource availability for other users and the environment?

- Appropriate **abstraction charges** should be levied on all users, in line with the "beneficiary pays" principle. They need to reflect at least the full cost of providing access to water resources. Proxies can be used to estimate the negative externalities associated with water abstractions so that they can also be reflected in abstraction charges. In cases of non-renewable water resources (such as certain groundwater resources exploited beyond sustainable yield), a charge reflecting the scarcity value of the resource could also be applied, reflecting the trade-off between mining water now or in the future.

Check 13. Are obligations related to return flows and discharges properly specified and enforced?

- Water entitlements need to be specified in a way that defines the **"net" amount of water** consumed, rather than the "gross" amount of water abstracted. In practice, there are numerous technical challenges that make it difficult (if not impossible) to measure net consumptive use with precision. However, rules of thumb can be applied to provide an estimation of net consumption according to the type of use. This approach can be used to maintain the integrity of the allocation regime, even while efficiency of use increases. There are alternative options to managing this issue (see discussion in Chapter 2, and example in Box 2.7).

- In addition to defining returns flow requirements from a quantitative perspective, the quality requirements (including thermal changes) for **discharges** also need to be clearly specified (see Box 2.8).

Check 14. Does the system allow water users to reallocate water among themselves to improve the allocative efficiency of the regime?

- Once the elements of a robust allocation regime are in place (including the role of government as the steward for the environment, as specified above), allowing water entitlement holders to **trade, lease or transfer** water entitlements (long term) and allocations within a given season (short term) can improve efficiency in allocation and resource use. It can also provide the conditions for improved management of risk of shortage, as water users have better incentives to manage risk and greater capacity to adjust to scarcity conditions. To avoid potentially negative impacts of trading arising from changing the location of water use, water entitlements and trading arrangements must be consistent with the overall limits of the resource. Where the trade, lease or transfer of water entitlements is possible, clear rules should be in place to facilitate transactions. Voluntary forfeiture of un-used water entitlements should be provided for.

- **Transaction costs** related to trading, leasing or transferring water entitlement and allocations should be kept **as low as possible**. This requires limiting trading costs to administrative costs that are unavoidable and also limiting third party interference in individual transactions.

Notes

1. This "Health Check" draws significantly on the background paper prepared for this project by Mike Young (2013), *Improving Water Entitlement and Allocation*.

2. Adaptive efficiency addresses the least cost path to maximise social welfare over the long term in the context of complex resources, unpredictability, feedback effects and path dependencies (Marshall, 2005).

3. "Latent demand" refers to demand reflected by entitlements that have been granted, but are not currently being exercising to take water at present. "Active demand" refers to demand reflected by entitlements that have been granted and in active use.

4. The agreement was signed by Spain and Portugal in 1998 and subsequently modified to comply with the EU WFD requirements on minimum annual flow for international rivers.

5. The management of exceptional circumstances, such as drought, is an art in itself, and merits a more in-depth examination than can be provided within the context of this *"Health Check" for Water Resources Allocation*.

References

Marshall, G.R. (2005), *Economics for Collaborative Environmental Management: Regenerating the Commons*, Earthscan, London.

OECD (2015), *Groundwater Use in Agriculture in OECD Countries: From Economic Challenges to Policy Solutions*, OECD Publishing, Paris, forthcoming.

OECD (2013a), *Water Security for Better Lives*, OECD Studies on Water, OECD Publishing, Paris, *http://dx.doi.org/10.1787/9789264202405-en*.

OECD (2013b), *Water and Climate Change Adaptation: Policies to Navigate Uncharted Waters*, OECD Studies on Water, OECD Publishing, Paris, *http://dx.doi.org/10.1787/9789264200449-en*.

South East Natural Resources Management Board (2011), "Guide to the development and contents of the 2009 Padthaway Water Allocation Plan", South East Natural Resources Management Board.

Young, M. (2013), *Improving Water Entitlement and Allocation*, background paper for the OECD project on water resources allocation (unpublished).

Glossary

Abstraction: The capture, diversion, taking of water for any purpose including an environmental purpose.

Allocation: An amount of water that an entitlement holder has been granted permission to abstract within a specified time period in a manner that is in accordance with pre-specified conditions.

Allocation regime: The combination of policies, mechanisms, and governance arrangements (entitlements, licenses, permits, etc.) used to determine who is allowed to abstract water from a resource pool, how much may be taken and when, as well as how much must be returned (of what quality), and the conditions associated with the use of this water.

Consumptive use: Water abstracted from a source such as a river, lake or aquifer, which is no longer available for use because it has evaporated, transpired, been incorporated into products and crops, or consumed by man or livestock.

Environmental flows (e-flows): The quantity, quality and timing of water flows required to sustain the ecological health of a water body.

Groundwater system: A connected body of water located beneath the earth's surface in soil pore spaces and/or in the fractures of rock formations.

Non-consumptive use: Water uses that do not substantially deplete water supplies, including swimming, boating, water skiing, fishing, maintenance of stream related fish and wildlife habitat, and hydropower generation.

Over-allocated: A water body with entitlements which if fully exercised would result in a level of abstraction that is greater than that which can be sustained.

Over-used: A water body where the quantity being abstracted is greater than that which can be sustained. See water stress.

Prior appropriation: A legal doctrine where the interests of the first person in time to take a quantity of water from a water source for a beneficial (agricultural, industrial and household) use has the right to continue to use that quantity of water for the same or similar purposes. Typically, this right is transferred along with the land associated with the use of this water and/or can be sold to others able to use it for a beneficial use without change in its priority status. The next user in time obtains a similar right provided its actions do not impinge on the rights of those with a more senior right. This is also known as a seniority regime, as the interests of most senior entitlement holders are met before any water is allocated to a more junior entitlement holder.

Public trust doctrine: Is a common-law principle of property law, held by states in some countries such as the United States, which establishes public rights in navigable waters and on their shores for the benefit of the public. The government is hence required to preserve them for the public's reasonable use, mainly for food, travel and commerce.

Resource pool: A water resource that can be managed as a single entity by issuing entitlements that are similar in form. Within a pool, all allocations are defined usually in a similar manner. In some cases, the pool is described as a management zone.

Return flow: The water physically withdrawn from a system and returned back to the same or a different water body following use.

Riparian entitlements: Under the riparian principle, all landowners whose property adjoins a body of water have the right to make reasonable use of it. If there is not enough water to satisfy all users, allotments are generally fixed in proportion to frontage on the water source. These rights cannot be sold or transferred other than with the frontage and water cannot be transferred out of the watershed.

Surface water: All water naturally open to the atmosphere, including rivers, lakes, reservoirs, streams, impoundments, seas or estuaries. The term also refers to springs, wells or other collectors of water that are directly influenced by surface waters.

Water body: A collection of connected water resource pools that can be managed as a single entity.

Water entitlement: The entitlement to abstract and use water from a specified resource pool as defined in the relevant water plan. In some countries, this may be referred to as "water rights", "water users' rights", "water contracts", abstraction license or permit.

Water scarcity: An imbalance between the supply and demand of freshwater as a result of a high level of demand compared to available supply, under prevailing institutional arrangements (including price) and infrastructural conditions.

Water stress: A measure of the total, annual average water demand of a river basin (or sub-basin) compared with the annual average water available (precipitation minus evapotranspiration) in that basin. Typically, these are grouped into four categories: less than 10% = no stress; 10-20% = low stress; 20-40% = medium stress; and more than 40% = severe stress.

ANNEX A

Questionnaire for the OECD project on water resources allocation

Overview

This questionnaire aims to gather an information base on *allocation regimes** in OECD and BRIICS countries, and in Argentina, Colombia and Costa Rica, to inform the OECD project on water resources allocation. The information collected will be used to draw out general trends and lessons associated with the design and functioning of water allocation regimes. It will also be used to develop "country profiles" to summarise the key elements of allocation regimes in a given country.

The questionnaire consists of two parts. Part 1 covers general contextual information at the national level and should only be completed once for each country. Part 2 covers specific elements of an example of an allocation regime. In many countries, there exist a number of different allocation regimes. For instance, allocation regimes may differ from one province/state/river basin to another. Allocation may also differ for surface water and groundwater systems.

If allocation of both surface and groundwater resources is dealt with in a uniform way across your country (there is only one allocation regime), Part 2 need only be completed once. Otherwise, please complete Part 2 two to three times, each time for a different example (groundwater system, surface water in a specific river basin, etc.). The choice of which water bodies to use as examples requires careful consideration. Please do not hesitate to discuss your selection of examples with the Secretariat.

Given the complexity of water resources allocation regimes, the questionnaire may require input from a number of experts with different types of expertise such as, legal, regulatory, planning and policy expertise. Please do not hesitate to circulate the questionnaire to the relevant experts in your country. For instance, an official of the national government would be well-placed to complete Part 1, whereas experts or government officials with detailed knowledge of specific examples of allocation regimes should respond to Part 2.

---

* Terms in bold and italics are defined in the appended glossary.

Questionnaire overview

Part 1. General contextual information

Section 1.1. Contextual information (institutional and legal setting; information on scarcity)

Section 1.2. Recent and current reforms of allocation regimes

Part 2. Specific information on an example of an allocation regime (2-3 examples from your country, would be welcome)

Section 2.1. Background information for the example of the allocation regime

Section 2.2. How the available resource pool is defined

Section 2.3. How users access water

Section 2.4. How access to water works in practice

Section 2.5. How exceptional circumstances (e.g. unplanned events or shocks negatively impacting the resource) are managed

Section 2.6. How access is monitored and enforced

Part 1. General contextual information

This part should be completed once per country in order to provide the institutional and legal context within which water allocation regimes operate and signal any recent or current reforms of allocation regimes. We recommend that this section be completed by an expert with a thorough knowledge of national policies and associated water allocation regimes.

Section 1.1. Contextual information (institutional and legal setting; information on scarcity)

1.1.1. Please list the institutions that are primarily responsible for water *allocation* (for example, Ministries, Water/River Basin Agencies, Water Users Association):

Institution	Scale of governance	Main responsibilities (i.e. policy, planning, issuing entitlements, monitoring and enforcement, etc.)
	Choose an item.	
	Choose an item.	
	Choose an item.	
	Choose an item.	
	Choose an item.	

1.1.2. What is the basis for your country's legal system as it pertains to water resources allocation?

❑ Common Law

❑ Roman/Statutory Law

❑ Other, specify: ____

1.1.3. How is the ownership of water resources legally defined (if at all)? *(Note: "Ownership" here refers to ownership of the resource itself, not the entitlement or right to use the resource.)*

a) **Groundwater:**

❏ Public ownership (e.g. **Public Trust Doctrine**)

❏ Private ownership

❏ Other, specify: ____

b) **Surface water:**

❏ Public ownership (e.g. Public Trust Doctrine)

❏ Private ownership

❏ Other, specify: ____

1.1.4. Has a mapping exercise been undertaken at the national level to identify areas where water scarcity is becoming a problem?

Groundwater: Yes ❏   No ❏

Surface water: Yes ❏   No ❏

a) If yes, please provide most recent reference(s): _______________________________

Section 1.2. Recent and current reforms of allocation regimes

1.2.1. Have any significant reforms of allocation regimes taken place in the last ten years?

Yes ❏   No ❏

a) If yes, please describe the process briefly and indicate the main driver(s) of the reform(s):

____

____

❏ Concerns about water shortages or scarcity

❏ Concerns about deteriorating water quality

❏ Concerns about equity in access to water

❏ Climate change

❏ Economic development

❏ Environmental improvement or protection

❏ Other, please indicate: _______________________________________

1.2.2. Is an allocation reform process underway or being considered in any part of your country?

Yes ☐   No ☐

a) If yes, please describe the process briefly and indicate the main driver(s) of the reform(s):

____

____

☐ Concerns about water shortages or scarcity

☐ Concerns about deteriorating water quality

☐ Concerns about equity in access to water

☐ Climate change

☐ Economic development

☐ Environmental improvement or protection

☐ Other, specify: _______________________________________

b) Please supply a link to or a copy of the most relevant documents outlining the nature of the reforms under consideration:

____

Part 2. Specific information on an example of an allocation regime (2-3 examples from your country, would be welcome)

This part seeks detailed information about a specific example of a water allocation regime. We recommend that it be completed by an expert familiar with allocation and governance arrangements for that particular allocation regime. In countries where there are a number of different water allocation regimes (for example, different allocation regimes for surface or groundwater, or variations in allocation regimes from one province/state/river basin to another), we recommend that Part 2 be completed 2-3 times for each example.

Section 2.1. Background information for the example of the allocation regime

2.1.1. Please indicate the name of the example of the allocation regime [for example, "The Incomati River Basin, South Africa", "The Colorado River Basin, USA", "The Province of Alberta, Canada", "Surface water systems in Korea (referred to under the River Act)"]:

____

2.1.2. Please indicate the territory/scale (e.g. river basin or catchment; state or provincial level; multi-purpose infrastructure) to which the responses in Part 2 apply:

____

2.1.3. Provide a brief (2-3 line) description of the physical characteristics of the water resource covered in this section of this example (e.g. variability of flow, nature of infrastructure if any, links between surface water and groundwater systems, if any):

____

____

____

2.1.4. To what extent can the flow rate of the water systems be managed or controlled?

The water system is:

❐ Fully regulated (the flow rate can be controlled fully)

❐ Partially regulated (the flow rate can be controlled to some extent)

❐ Not regulated (the flow rate cannot be controlled)

2.1.5. Provide a *general* estimation of the % of mean annual inflow/recharge that is consumed (% agriculture, % urban, etc.), if available:

Water uses	Percentages
Agriculture	
Domestic	
Industrial	
Energy production (not including hydro power)	
Environment (evapotranspiration)	
Transfer to the sea or another system	
Other, specify:	
Total	**100%**

a) Please indicate any significant **non-consumptive uses** (e.g. hydro power, transport, etc.) in this water system, if any:

____

Section 2.2. How the available resource pool is defined

2.2.1. Is there a clear definition of the limit on **consumptive use**?

❐ There is a limit in the volume of water that can be abstracted

❐ There is a limit to the proportion (e.g. percentage) of water that can be abstracted

❐ There are restrictions on who can abstract the water (but no limit on how much water can be abstracted)

❐ There is no explicit limit on water abstraction

a) Is the amount of water available for consumptive use in the resource pool linked to a public planning document? (E.g. a river basin management plan.)

❐ Yes, the limit is linked to a river basin management plan

❐ Yes, the limit is linked to another planning document, please indicate: ____________

❐ No, the limit is not linked to any planning document

If yes,

b) Who is the authority responsible for preparing the planning document?

____

c) What is the nature of the plan?

❐ Statutory instrument that must be followed

❐ Guiding document

2.2.2. Are minimum *environmental flows (e-flows)*/sustainable diversion limits defined?

Yes ❑   No ❑

a) If yes, provide detail on how e-flows are defined: _________________________

____

b) Is freshwater biodiversity taken into account? Yes ❑   No ❑

If so, how? ____

c) Is terrestrial biodiversity taken into account? Yes ❑   No ❑

If so, how? ____

2.2.3. Are the following factors taken into account in the definition of the available resource pool?

Factor	Not taken into account	Taken into account	If taken into account, how?
Non-consumptive uses (e.g. navigation, hydroelectricity)			
Base flow requirements			
Return flows (how much water should be returned to the resource pool, after use)			
Inter-annual and inter-seasonal variability			
Connectivity with other water bodies			
Climate change			

2.2.4. Is the water system currently considered:

❑ **Over-allocated** (e.g. current use is within sustainable limits but there would be a problem if all legally approved entitlements to abstract water were used)

❑ **Over-used** (existing **abstractions** exceed the estimated proportion of the resource that can be taken on a sustainable basis)

❑ Neither over-allocated nor over-used

2.2.5. If the water system is currently over-allocated or over-used, please indicate which measures are being undertaken to address this issue (if any)?

____

2.2.6. What arrangements are in place, if any, to accommodate the potentially adverse impacts of climate change on the resource pool? (E.g. using best available science to plan for future changes in availability, undertaking periodic monitoring and updating of available pool.)

____

Section 2.3. How users access water

2.3.1. Are private entitlements to take water defined? Yes ❑   No ❑

a) If yes, is it in the form of:

❑ An individual entitlement (to an individual person)

❑ A collective entitlement: (to a group of persons/organisation/city)

❑ Other, specify: ____

i) If there are collective entitlements they are assigned to:

❑ An institution representing water users (e.g. WUAs)

❑ Another (perhaps informal) community-based arrangement

❑ Other, specify: ____

ii) In the case of collective entitlements, please indicate the process for allocating water among individual users within a group of users:

❑ Bargaining process

❑ Informal trading

❑ Other, specify: ____

b) If private entitlements are not allowed, please describe the constraints (e.g. public management is seen to be needed for equity, social concerns, etc.):

____

2.3.2. Are water users' *entitlements* to abstract or divert water from the resource pool legally defined? *(Note: Depending on the country context, "water users' entitlements" may also be referred to as "water users' rights" or "abstraction licenses or permits". Note that this section refers to entitlements to use the resource, not ownership of the resource itself.)*

Yes ❑   No ❑   In process of development ❑

a) If yes:

i) What is the nature of the water users' entitlements?

❑ **Water entitlements** unbundled from property titles

❑ **Riparian entitlements**

❑ **Prior appropriation** where reliability is a function of the year when the entitlement was first issued

❑ Other, specify: ____

ii) How are entitlements defined?

❑ Purpose that water may be used for

❑ Maximum area that may be irrigated

❑ Maximum volume that may be taken in a nominated period

❑ Proportion of any water allocated to a defined resource pool

❑ Other, specify: ____

iii) Typically, how long are entitlements issued for?

    ❐ A term of years ______ without expectation of renewal

    ❐ A term of years ______ with expectation of periodic renewal

    ❐ In perpetuity but conditional upon beneficial use

    ❐ In perpetuity

    ❐ Other, specify: ____

iv) Are return flow obligations specified?

    Yes ❐ No ❐

    If yes, how? (E.g. % of the entitlement.) _______________________

b) If these entitlements are *not* legally defined, please explain how abstraction and use works in practice:

____

2.3.3. What types of users are **not** required to hold a water entitlement but can still take water from the resource pool? (E.g. livestock, domestic users, urban water suppliers, farm dam owners, small scale users, people who plant trees?)

____

a) Please provide an estimate (if available) of the percentage of total water uses related to these groups of users?

____

b) How are the adverse impacts of any increase in these uses controlled?

____

2.3.4. Is there a pre-defined set of priority uses within the resource pool? Yes ❐ No ❐

a) If yes, please indicate the sequence of priority uses below:

Categories	Sequence of priority (provide further details if necessary)
Agriculture	
Domestic	
Industrial	
Energy production	
Environment	
Transfer to the sea or another system	
National security (e.g. protection of infrastructure and critical dikes, nuclear plants)	
Other, specify:	

2.3.5. Do users pay abstraction charges?

Categories	Yes/no	Basis for charge [i.e. volumetric (metered), proxy (e.g. surface of irrigated land), other (specify)]	Does pricing instruments reflect water scarcity? (Yes/no)
Agriculture			
Domestic			
Industrial			
Energy production (not including hydro power)			
Hydro power			
Other, specify:			

a) If pricing arrangements reflect scarcity, briefly indicate how:

____

Section 2.4. How access to water works in practice

2.4.1. If there are new entrants and/if entitlement holders want to increase the volume of water they use in the resource pool, can new entitlements be issued or existing entitlements be augmented?

- ❐ Yes, without restriction
- ❐ No, catchment is closed
- ❐ Conditional on:
    - ❐ Assessment of third party impacts
    - ❐ Environmental impact assessment (EIA)
    - ❐ Existing user(s) forgoing use
    - ❐ Other, specify: _______________________________

2.4.2. Characteristics of entitlements:

a) If the entitlement is not used in a given period, the entitlement:

- ❐ Will be lost (e.g. "use it or lose it")
- ❐ Remain in place for the period it is issued for
- ❐ Other, specify: _______________________________

b) Are entitlements differentiated based on the level of security of supply (or risk of shortage)? Yes ❐   No ❐

If yes, how? _______________________________________

c) Can water users' entitlements be traded, leased, transferred in any way? Yes ❐   No ❐

If users' entitlements can be traded, leased, transferred, how does this work in practice?

____

d) Are allocations (the amount that can be taken at any point in time) managed separately from entitlements? Yes ❐   No ❐

e) Is allocation trading allowed? Yes ☐   No ☐

   i) If yes, how is the price at which water is traded determined?

____

   ii) What administrative charges are associated with an allocation trade?

____

   iii) Are there specific restrictions on trading of entitlements and/or allocations?
Yes ☐   No ☐

   If yes, what are the restrictions? (E.g. by certain user groups, in certain locations, etc.)

____

____

f) Can the entitlement function as a financial instrument? If so, how?

____

Section 2.5. How exceptional circumstances (e.g. unplanned events or shocks negatively impacting the resource) are managed

2.5.1. How is the amount of water made available for allocation varied from time to time? (E.g. from year to year, between seasons, etc.)

____

2.5.2. Is there a distinction between the allocation regimes used in "normal" times and in times of "extreme/severe" water shortage? Yes ☐   No ☐

a) If yes, how are "exceptional" circumstances defined? (E.g. extended drought, etc.)

____

b) How does this affect the allocation regime? (E.g. triggers the water use restrictions, reduction in allocations according to a pre-defined priority uses, suspension of the regime plan, etc.)

____

____

c) Who is the responsible authority for declaring the onset of "exceptional" circumstances?

____

d) Are stakeholders involved in the definition of "exceptional" circumstances? Yes ☐   No ☐

   If yes, how? ____

Section 2.6. How access is monitored and enforced

2.6.1. Please indicate if withdrawals are monitored:

Categories	Monitored? (Yes/no) If yes, how? (I.e. metering, aerial surveillance, or other, please specify)	Who is the responsible authority?	Type of sanction(s) for non-compliance, if any
Agriculture			
Domestic			
Industrial			
Energy production			
Environment			
Transfer to the sea or another system			
National security (e.g. protection of infrastructure and critical dikes, nuclear plants)			
Other, specify:			

2.6.2. Are there any type of conflict resolution mechanisms in place? Yes ❐   No ❐

a) If yes, briefly describe them and indicate which institutions are involved:

Conclusion: Other information

Is there any else that you consider that the Secretariat should be aware of as it interprets the information supplied in this response? We would be particularly grateful for copies of any documents that provide a short description or review of the regimes described in Part 2.

ORGANISATION FOR ECONOMIC CO-OPERATION AND DEVELOPMENT

The OECD is a unique forum where governments work together to address the economic, social and environmental challenges of globalisation. The OECD is also at the forefront of efforts to understand and to help governments respond to new developments and concerns, such as corporate governance, the information economy and the challenges of an ageing population. The Organisation provides a setting where governments can compare policy experiences, seek answers to common problems, identify good practice and work to co-ordinate domestic and international policies.

The OECD member countries are: Australia, Austria, Belgium, Canada, Chile, the Czech Republic, Denmark, Estonia, Finland, France, Germany, Greece, Hungary, Iceland, Ireland, Israel, Italy, Japan, Korea, Luxembourg, Mexico, the Netherlands, New Zealand, Norway, Poland, Portugal, the Slovak Republic, Slovenia, Spain, Sweden, Switzerland, Turkey, the United Kingdom and the United States. The European Commission takes part in the work of the OECD.

OECD Publishing disseminates widely the results of the Organisation's statistics gathering and research on economic, social and environmental issues, as well as the conventions, guidelines and standards agreed by its members.

OECD PUBLISHING, 2, rue André-Pascal, 75775 PARIS CEDEX 16
(97 2015 04 1 P) ISBN 978-92-64-22962-4 – 2015

IWA Publishing's authorised EU representative for General Product Safety Regulations is Diane D'Arras, 15 rue Duret, 75116 Paris, France, e-mail: safety@iwap.co.uk.

Printed and bound by CPI Group (UK) Ltd, Croydon, CR0 4YY
12/05/2026
02108893-0011